Home Automation Basics

Practical Applications Using Visual Basic® 6

by Thomas E. Leonik, P.E.

Home Automation Basics

Practical Applications Using Visual Basic® 6

by Thomas E. Leonik, P.E.

PROMPT® Publications is an imprint of Sams™ Technical Publishing, 5436 W. 78th St., Indianapolis, IN 46268.

International Standard Book Number: 0-7906-1214-3

Library of Congress Catalog Card Number: 00-105991

Acquisitions Editor: Alice J. Tripp
Editor: Will Gurdian
Assistant Editor: Kim Heusel
Typesetting: Will Gurdian
Proofreaders: Pat Brady, Kim Heusel
Cover Design: Christy Pierce
Graphics Conversion: Christy Pierce, Bill Skinner
Illustrations and Other Materials: Courtesy Rockwell Automation Allen-Bradley

PRINTED IN THE UNITED STATES OF AMERICA

9 8 7 6 5 4 3 2

DEDICATION

To those awesome girls in my life:

my daughters, Kaitlyn Tara Leonik and Kallie Anne Leonik;

my wife, Daine Leonik;

and my sister, Bernadine Leonik

ACKNOWLEDGMENTS

I would like to start by thanking Microsoft® for making such a great product. You can't beat Visual Basic® 6 for rapid application development. Microsoft does not state or imply any certification or approval of the material covered in this book.

Special thanks to my local Allen-Bradley distributor, Fromm Electric, located in Cherry Hill, N.J., for providing me with an evaluation of the Allen-Bradley MicroLogix PLC with analog.

The Fromm Electric Crew:
Barry ("We'll-State-Contract-Everything") Emers – Drives
Tom Rosado – PLCs
Steve Bruno – General Sales

Special thanks to Rockwell Automation Allen-Bradley for making excellent PLCs, and for granting me the permission to discuss the MicroLogix PLC and the two communication commands Unprotected Read/Unprotected Write. Thanks again to Rockwell Automation for providing excellent photographs of the MicroLogix PLC. Rockwell Automation Allen-Bradley does not state or imply any certification or approval of the material covered in this book.

Special thanks also to Industrial Electronic Engineers, Inc. for sending alphanumeric display photographs and granting me permission to discuss communicating to an IEE serial display. Industrial Electronic Engineers, Inc. does not state or imply any certification or approval of the material covered in this book.

Finally, special thanks to my family for giving me the space to write this book, and to my faithful companion, Shadow, a true surf dog, who dragged me out every night for 2.5-mile walks—keeping me fit through the many months of midnight oil burning that it took to complete this project.

Contents

PREFACE

This book explores the world of Visual Basic 6 programming with respect to real-world interfacing, animation, and control on a beginner/intermediate level. Most Visual Basic books on the market today and in the past do a good job of describing various control elements of Visual Basic and how they work. Examples are provided for programming some type of database application. The focus of this book, however, is interfacing to an external device via the serial port, showing the status of this device by animating objects on a Visual Basic form, and then controlling this device. The Allen-Bradley MicroLogix PLC by Rockwell Automation is one serial device that will be explored. An alphanumeric display is another serial device that will be investigated.

The acronym "PLC" stands for Programmable Logic Controller. PLCs are the fundamental building blocks in industrial control systems today. Typically, a PLC consists of inputs, outputs, a central processing unit, user memory for control functions, a proprietary operating system, and a serial port. With a PLC, input devices such as push buttons and limit switches—to name a few—are wired into inputs, while output devices such as lights, horns, motor-control relays, and so on are wired to outputs. All cross wiring required to implement a particularly desired function is accomplished with software, rather than with hard wiring and additional components.

The aim of this book is Visual Basic, but it also will discuss to a certain degree the architecture and programming of PLCs. The most popular language used to program PLCs is called Ladder Logic. Ladder Logic is modeled on the way one would actually wire up devices to relays. Input devices are represented as contacts, and outputs are represented as relay coils. Essentially, each rung of Ladder Logic is a graphical form of a Visual Basic "if-then" statement. Typically, the PLC instruction set supports internal timers, counters, math functions (integer and real via floating point), move functions, Boolean functions, and communication functions. The Rockwell Automation MicroLogix PLC was selected because it is a popular, powerful, and inexpensive PLC. Most of the material presented on the MicroLogix PLC will also apply to other manufactured PLCs.

The application example presented in this book will be that of using the PLC as a home monitor. This home-monitor example is more appealing and easier to understand than one of a wastewater pumping station with level control using a variable frequency drive. Some day, the PLC or some form of it will be controlling all the electrical operations of a house.

As a home monitor, the PLC will observe the following digital inputs: front doorbell push button, rear doorbell push button, front door-open sensor, rear door-open sensor, first-floor HVAC system on, second-floor HVAC system on, water pump on, and mailbox-open sensor. In addition to the discrete digital inputs, five temperatures will be monitored (outside air, crawl space, first floor, second floor, and great room), as will the water pressure of the water pump. Two PLC outputs will be used to drive a horn and turn on a blower fan under the control of the Visual Basic program.

A Visual Basic 6 program will be developed that communicates serially with the PLC using the Rockwell Automation Allen-Bradley protocol. This protocol will be considered to perform two tasks: an "unprotected read" and an "unprotected write." The structure of this binary protocol will be examined. The software developed to perform these tasks will be simply and clearly presented.

Another Visual Basic 6 program will be developed that interprets collected information and animates objects on a Visual Basic form accordingly, such as a door shown in the open or closed position along with an associated time/date stamp as to when it was opened or closed. Temperature and other data information will be stored as monthly files on the hard drive in CSV (comma-separated values) format. Microsoft Excel can then be used to provide a graphical representation of the data contained in these files. Wave files will be launched to provide audio feedback of certain events.

A stand-alone, full-function graphing program will be developed that interprets the data log files. This program will automatically update in conjunction with the data-log file.

A second serial port will be used to provide textual-based status information from the Visual Basic home-monitor program to a remote alphanumeric display.

Through the process of developing the home-monitor program, many aspects of Visual Basic programming will be studied. Emphasis will be placed on keeping the software as simple as possible. The lessons learned in this book will be invaluable for future serial and animations projects.

All programs presented in this book and contained on the companion CD-ROM have been thoroughly tested.

ABOUT THE AUTHOR

Thomas E. Leonik is an electrical engineering graduate of Villanova University and a licensed Professional Engineer in the state of New Jersey. After graduating, Leonik worked in the design, fabrication, and testing of CMOS semiconductor-integrated circuits for several years and rose to the position of lead engineer for timekeeping and custom circuits.

Eager to learn the control aspects of his field, he moved on to work as a design engineer for a prominent process-control manufacturer. Leonik worked on various closed-loop control designs, data acquisition, and interface circuits. These circuits were implemented using both discrete digital logic circuits and microprocessors.

Currently, Leonik works as a consultant in the areas of industrial controls and automation, microprocessor design, data acquisition, and specialized software. He is proficient in assembly-language programming for several types of microprocessors. Microprocessor designs are taken from concept to reality in the form of a finished printed circuit board. Typically, the PC interface for these specific designs is written in either Visual Basic or Visual C++. Leonik is also proficient in ladder logic programming for several types of Programmable Logic Controllers (PLCs) used in industrial controls and automation.

To date, Leonik has provided complete industrial solutions for several wastewater treatment facilities, including the corresponding remote pumping lift stations, an in-vessel sludge composting facility, and a landfill leachate pumping system. In the past, the PC animation for these various projects was accomplished with commercially available software designed specifically for the animation task. The most recent animation projects have been written exclusively in Visual Basic. does not state or imply any certification or approval of the material covered in this book. Visual Basic is an excellent platform for the animation task because of its power, ease of use, and Windows-based orientation.

Home Automation Basics

Practical Applications Using Visual Basic® 6

Chapter 1
VISUAL BASIC BASICS

INTRODUCTION

The purpose of this chapter is to provide a whirlwind overview of Visual Basic 6. The world of Visual Basic has evolved to the point that it's impossible to adequately cover with just one book; truly, it would take volumes. If this book is your first exposure to Visual Basic 6, then you should seek the aid of other beginner books and carefully study the Help files and tutorials Microsoft provides. If you are a somewhat seasoned Visual Basic programmer, you may want to just quickly scan this chapter or jump right into the next chapter.

Visual Basic is an event-driven, object-based structured programming language. This description may at first make Visual Basic sound a little intimidating, but its approach to programming is actually very logical and intuitive.

The term "visual" is applied to the name due to the graphical nature of placing, rearranging, and accessing objects in the development mode. In Visual Basic, everything is essentially object-based. You start out with an object to contain your desired application. This container happens to be a window into the operating system. In the program-development environment, this container is called a **form**. You place the objects required for a specific programming task into this form. These objects are called **controls**. Command button, timer, label, shape, and textbox are some examples of control objects.

In general, all objects consist of properties or characteristics that can be accessed either immediately in the development mode or while the application is running with Visual Basic code. Similarly, an object may have events associated with it. Double-clicking the mouse on an object or moving the mouse over an object are examples of events.

When an event occurs, software specific to the event is executed. The event-driven nature of Visual Basic programming is a radical departure from the more traditional sequential flow of early Basic programming, so care must be taken to ensure that the proper sequence of events has occurred to achieve the desired results.

Once you've gained some experience in developing applications with Visual Basic, it's hard not to appreciate the rapid application development that is possible. Basic has come a long, long way. Great job, Bill Gates and the Microsoft Visual Basic development team!

Getting Started

Once you've installed Visual Basic 6 on your hard disk, there are a number of ways to start it. It's directly available by hitting the Windows Start Button on your keyboard, or by clicking the Start Button on the task bar with your mouse and navigating to the Programs category with either the cursor keys or the mouse. Look for Visual Basic 6. This is illustrated in Figure 1.01. (Keep in mind that this illustration is showing my computer desktop with the programs I have loaded; the way it looks on your machine is a function of the programs you have loaded.) The most common display resolution setting used in this book is 1024 x 768.

I always create a shortcut for the programs that I access frequently and place it in the Quick Launch toolbar. You can easily accomplish this task by right-clicking on the VB6 icon from the Programs category and then dragging the icon to the desktop. As soon as you release the mouse button, a dialogue box appears asking if you want to move, copy, or create a shortcut. In this case, you want to create a shortcut. Then grab the shortcut and drop it into the Quick Launch area of the task bar. On the other hand, you may prefer to just leave the shortcut on the desktop. Figure 1.02 shows the location of the Quick Launch toolbar.

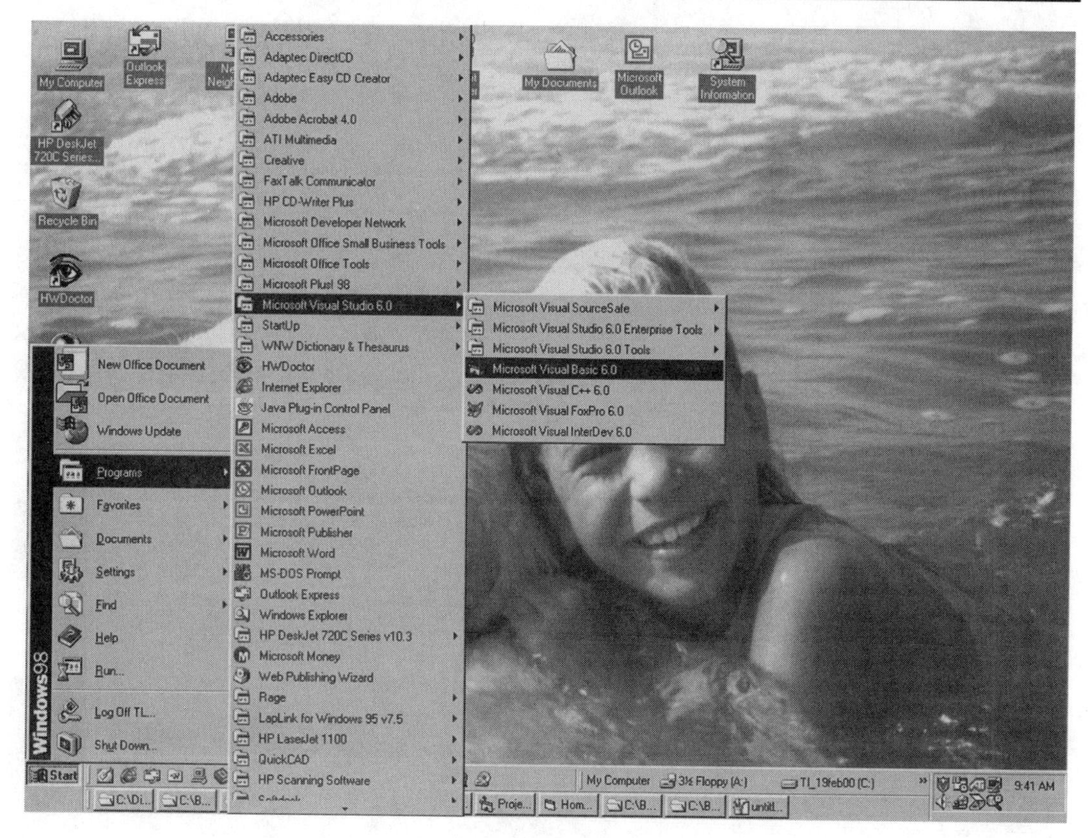

Figure 1.01

Once you have launched Visual Basic 6, the New Project dialogue window (as shown in Figure 1.03) will pop up to greet you. You have your choice between New, Existing, and Recent. This feature can be disabled in future runs of Visual Basic by clicking on the "Don't show ..." checkbox in the bottom left-hand corner.

Integrated Design Environment

Figure 1.04 shows the elements of Visual Basic 6 design environment. The Menu bar holds the commands needed to build an application. You should navigate through the different pull-down menus to get acquainted with the available features. Help is located on the Menu bar. Selecting an entity and pressing F1 will also launch Help.

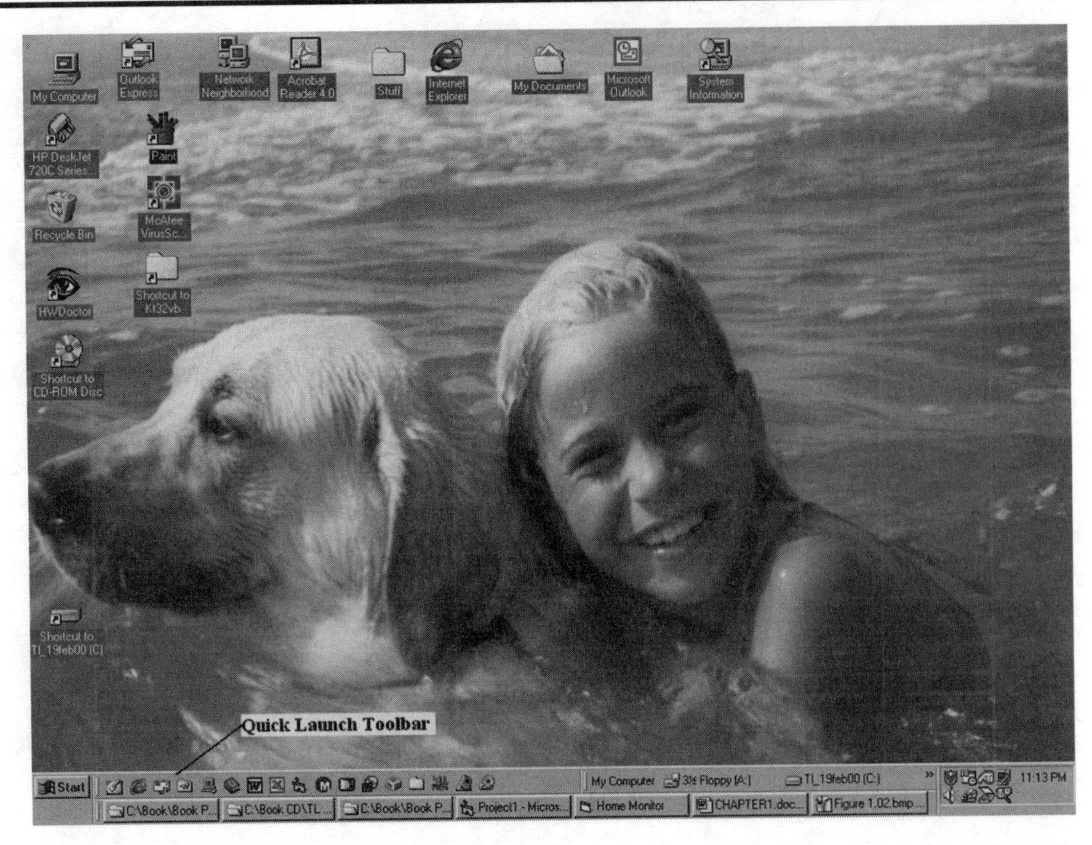

Figure 1.02

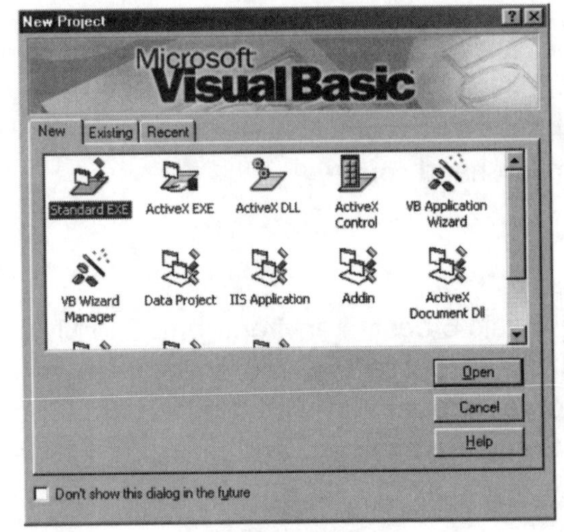

Figure 1.03

The Toolbar provides quick access to frequently used commands. If you position the mouse cursor over any of the icons on this window, a pop-up descriptive text window will appear. This text window is called a ToolTipText window.

The Toolbox holds the various controls for any project. Shown in Figure 1.04 are the standard default controls that appear when you start a new project. Other controls can be added to the Toolbox. Many additional controls are

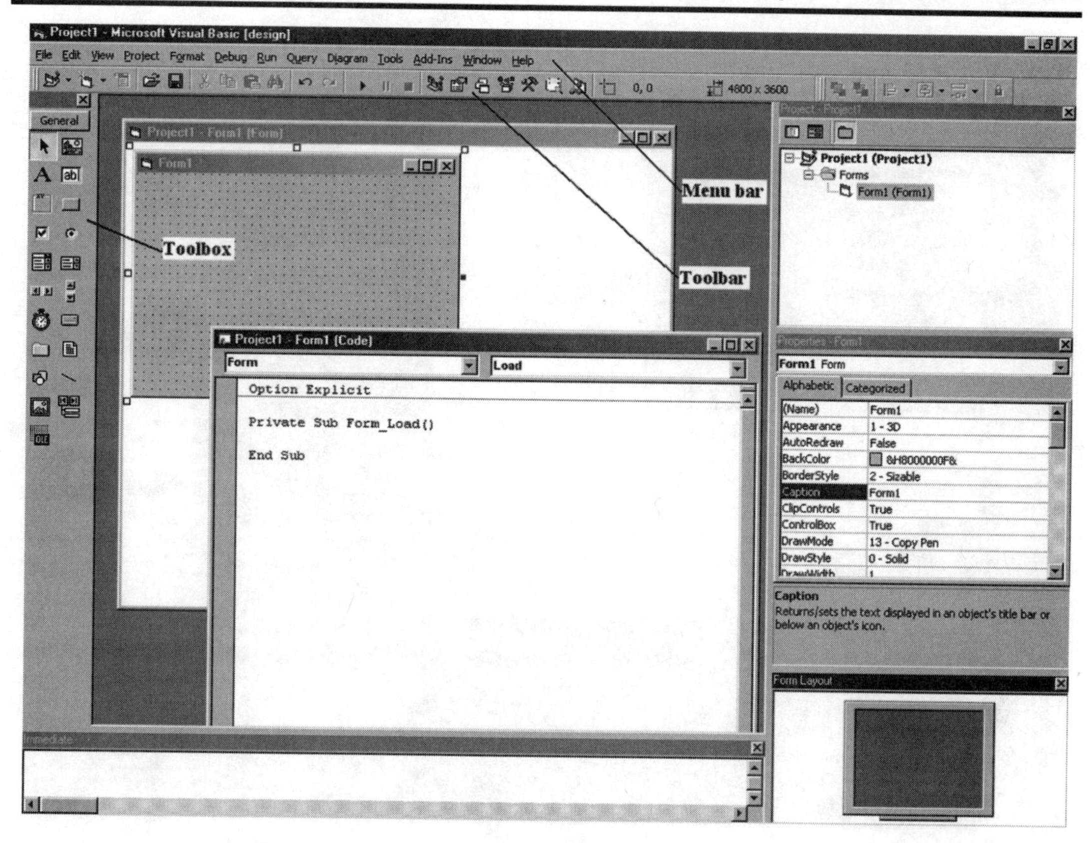

Figure 1.04

supplied with Visual Basic, and many others are available through third-party developers. To add any control not available in the Toolbox, you need to right-click on a background area of the box. At this point, a dialogue box will appear, as illustrated in Figure 1.05.

When you select "Components..." a new dialogue box opens, as illustrated in Figure 1.06. Scroll down and place a check mark on the desired control or controls. Follow the same procedure to remove a control from the Toolbox. In this case, remove the check mark from the unwanted control. The dialogue box in Figure 1.05 also reveals an optional tab feature that may be installed to increase the Toolbox capacity.

Each control is an object, and as such has properties. The property window will show the properties or characteristics of the control that is selected on the form.

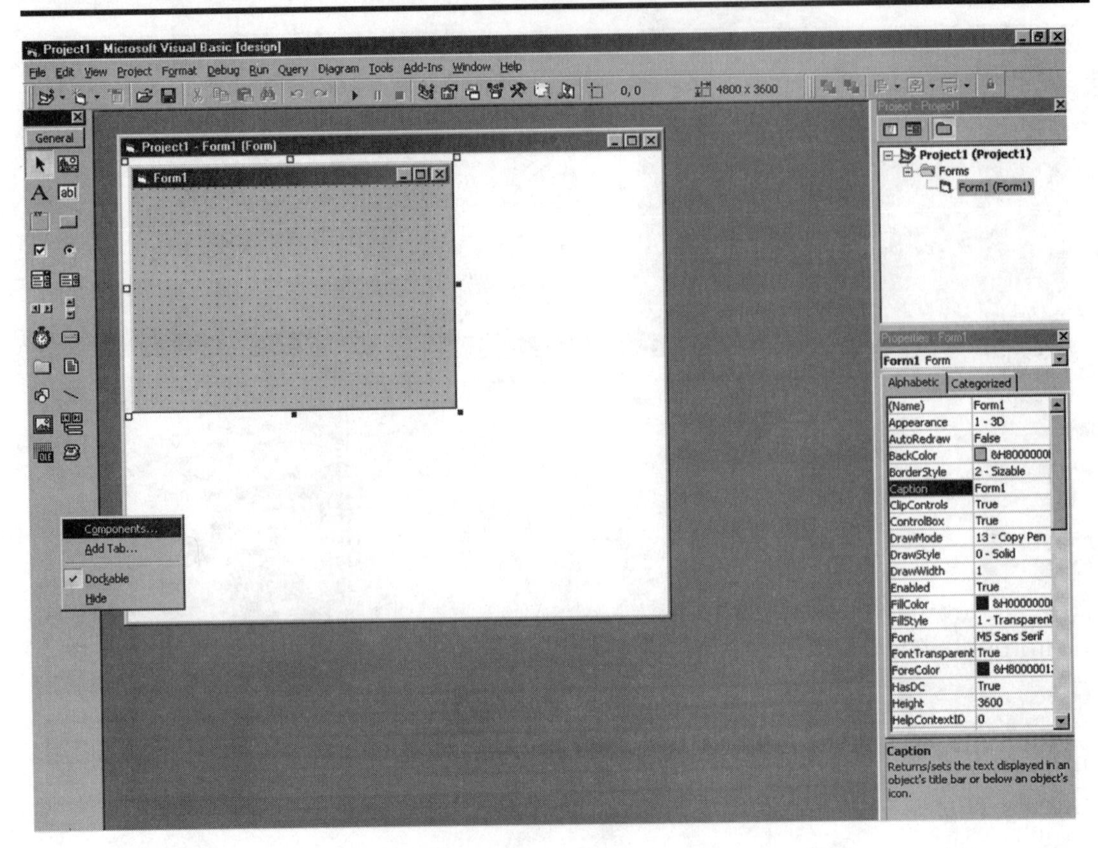

Figure 1.05

Some object properties are caption, color, font, and enabled. Any control property can be viewed from the property window by selecting it with the pull-down menu.

The Project window lists all the forms and modules contained in the project. Double-clicking on any form or module will bring it to focus. In the case of forms, there are selection buttons at the top of the window for viewing either the form or the code.

The form is the container that accepts the controls for a project. If you show the form, it will typically be the human interface window for your project. You may have a form that houses the controls for some particular task that requires the form to be hidden. The form can be adjusted in the design mode to any specific size. As with any object, the form has properties that affect the way it looks and behaves. With the Form Layout window, you can position the form to the specific location in which you want it to appear at run time.

If the Form (Code) window is not visible, you can always activate it simply by double-clicking on the respective Form (Form) window, or by right-clicking on the desired form in the Object Browser window and then selecting "View Code."

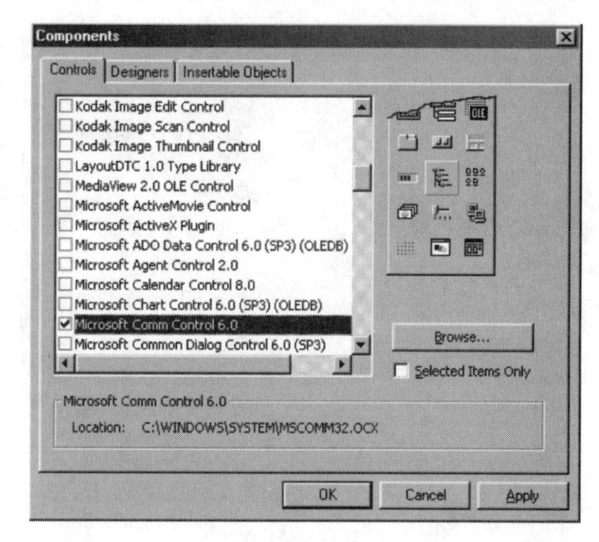

Figure 1.06

The Form (Code) window is where the code is placed. Any object placed on the form that has an event property will appear in the left list box of the Code window. The right list box will automatically show all possible event procedures for the selected object.

The Immediate window is useful for debugging purposes. Statements can be placed inside the code that prints out in this window during execution of the program. "Debug.Print" is the statement that would be placed in code, followed by the variable to be viewed. In addition, if the program is in break mode, the Immediate window can be used to view specific data by manually typing in a print request. Break mode is achieved by setting breakpoints in the code at particular locations. Breakpoints are either toggled on or toggled off through the Debug menu. When the program hits a breakpoint, Visual Basic halts execution of the program until instructed to proceed. All the data values are preserved when in break mode.

The Object Browser tool is extremely useful. This tool will provide a list of objects available for use in your project based upon the included components. In addition, you can use the Object Browser to explore objects in Visual Basic and determine what methods and properties are available for those objects. Click on the item to select it. Use function key F1 to launch Help specifically for that item. A few more clicks will take you to an example where—if desired—the code can be cut and pasted directly into your application.

The Visual Basic 6 integrated design environment is extremely accommodating. As code is written, the statements can be automatically checked for syntax, automatic object-member listing can be provided for quick statement-

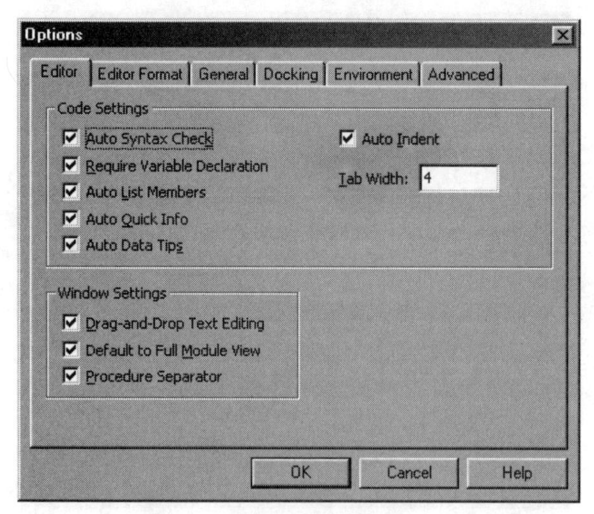

Figure 1.07

completion information, automatic quick information about functions and parameters can be enabled, and automatic data tips provide the contents of variables when the program is paused for debugging purposes.

Make certain that these functions are enabled by going to Menu bar and selecting Tools/Options/Editor tab. You should see a dialogue box like the one in Figure 1.07.

Variables

When you work in a programming language, you need places to store values as you go about performing a task. In Visual Basic, as well as in many other programming languages, these storage areas are called **variables**. Each variable is assigned a meaningful name and data type.

In addition to a single-storage data location, blocks of consecutive and indexable data-storage locations are possible. These blocks of contiguous data are called **arrays**. Arrays can also be multidimensional. Spreadsheets are structured like multidimensional arrays. Like an array, each cell in a spreadsheet is indexable.

Visual Basic supports many different data types. Table 1.01 illustrates the various data types, the amount of memory each type consumes, and the range of viable values for each type. Unlike some other programming languages, Visual Basic doesn't allow you to specify whether a data type such as byte, integer, and long is to be classified as signed or unsigned. For example, you may think that a number such as 40,000, which is below 2^16, may be represented by 2 bytes or by an integer value. Integers in Visual Basic are intrinsically signed. The largest possible positive integer number is 32,767. If you try to place 40,000 into an integer value, you will most certainly get an overflow error. There is, however, no problem with

Data type	Byte size	Range Low Limit	Range high Limit
Byte	1 byte	0 to 255	0 to 255
Boolean	2 bytes	FALSE	TRUE
Integer	2 bytes	-32,768	32,767
Long (long integer)	4 bytes	-2,147,483,648	+2147483647
Single (single-precision floating-point)	4 bytes	-3.402823E38 to –1.401298E-45 for negative values	1.401298E-45 to 3.402823E38 for positive values
Double (double-precision floating-point)	8 bytes	-1.79769313486231E308 to -4.94065645841247E-324 for negative values	4.94065645841247E-324 to 1.79769313486232E308 for positive values
Currency (scaled integer)	8 bytes	-922,337,203,685,477.00	922,337,203,685,477.00
Decimal	14 bytes	79,228,162,514,264,337,593,543,950,335 with no decimal point; -7.9228162514264337593543950335 with 28 places to the right of the decimal Smallest non-zero number is -0.0000000000000000000000000001	+79,228,162,514,264,337,593,543,950,335 with no decimal point; +7.9228162514264337593543950335 with 28 places to the right of the decimal Smallest non-zero number is +0.0000000000000000000000000001
Date	8 bytes	January 1, 100	December 31, 9999
Object	4 bytes	Any Object reference	
String (variable-length)	10 bytes + string length	0	Approximately 2 billion

Table 1.01

Continued on next page

String (fixed-length)	Length of string	1	Approximately 65,400
Variant (with numbers)	16 bytes	Any numeric value up to the range of a Double	
Variant (with characters)	22 bytes + string length	Same range as for variable-length String	
User-defined (using Type)	Number required by elements	The range of each element is the same as the range of its data type.	

Continued from previous page

placing 40,000 into a long integer data type. By studying Table 1.01 and by becoming familiar with the range of values for each data type, you can prevent future programming problems. According to Table 1.01, it sure looks like Visual Basic 6 will have a Y10K problem.

Declaration of variables is the process by which you name a variable and declare the data type. Although Visual Basic doesn't force you to explicitly declare a variable like other programming languages do, experience dictates that declaration of variables is well worth the effort in the long run. Ultimately, you save time by reducing the number of programming errors caused by typos. You can and should specify that variable declaration be required with an Option Explicit statement in the declaration section of each form. Additionally, you can specify that variable declaration be automatic with each project by enabling it at Menu/Tools/Options/Editor tab and placing a check mark on "Require Variable Declaration." With variable declaration enabled, Visual Basic will alert you to any undeclared variable in your code.

Another good programming technique is to implement a naming convention for variables, objects, and code modules. Essentially, a name must begin with a letter and must be unique. A name cannot exceed 255 characters, cannot be a keyword, and cannot contain an embedded period, space, or embedded type-declaration character (! @ # $ % &). The Help documentation does a thorough job of describing the recommended naming convention for the various entities. Supplemental naming-convention information is available at the Microsoft Developer's Web Site.

For future code readability, you should adopt some naming scheme and apply it consistently. With Visual Basic Code Editor, you can right-click on a variable name to retrieve additional information on the variable. Some of the choices available are "Quick Info" and "Definition." The Quick Info selection will balloon out with the data type. Quick Info is also accessible by keying Ctrl+I. The Definition selection will take you directly to the location in your project where the variable is declared. Visual Basic Code Editor will also attempt to complete the word that you are presently typing by activating "Complete Word." Either right-click on the mouse or key in Ctrl+space to activate the Complete Word function. This becomes especially useful for items with long significant names.

The most important point to assigning a name is to select one that is descriptive of the task at hand. In addition, you should vary the case of the name. For example, a variable name such as TheNumberOfBytesCollected is much more readable than thenumberofbytescollected. Visual Basic is not case sensitive; therefore, you only have to use the shift key to apply the upper case letters during the declaration process. When you are in the process of coding, type the name all in lower case letters. Code Editor will automatically change the case of the typed variable name at the completion of your statement to reflect the name given when the variable was declared. This feature provides you with visual feedback that you typed in the correct name. Otherwise, if you misspelled a variable name, you will get a "Compile Error: variable not defined" message when you run the code.

The variable naming convention used in this book is presented in Table 1.02. Using lower case prefix identifiers for the naming convention is commonly called Hungarian notation. Naming used for a numeric constant will follow suit, except all the letters in the name are upper case. Sometimes the word constant is included in the name in lower case characters. Frequently, in many of the programs presented later in this book, a single letter name will be used in the local procedure for indexing purposes, and the prefix will be omitted.

Data Type	Prefix	Example	Comment
Byte	b	bData1	
Boolean	Suffix flag	GotByteflag	Note suffix use
Integer	i	iData	
Long	l	lResult	
Single	f	fProduct	Use f for floating point
Double	df	dfProduct	df for double floating point
String	s	sText	
Object	o	oChart1	
Variant	v	vMessage	

Table 1.02

Control	Prefix	Example
CommandButton	cmd	cmdLoadFile
Label	lbl	lblMonth
OptionButton	opt	opt9600Baud
TextBox	tb	tbTransmittedBytes
CheckBox	cb	cbAutomaticTransmit
Timer	tmr	tmrUpDate
Shape	sh	shWaterTank

Table 1.03

The naming convention for the standard intrinsic controls used in this book is presented in Table 1.03. With respect to the programs presented in this book, in some cases, if the control is not really doing anything special, the Visual Basic-assigned name is used.

Keywords

If you are writing assembly code for a microcontroller chip such as an Intel 80C32 or a Microchip Pic family chip, you had better be familiar with the chip's architecture and instruction set. Visual Basic is a high-level language. It also has an instruction set, which is commonly referred to as **keywords**. Some of the primary keywords will be examined later in this chapter. You should always use Microsoft's Help files for additional information. These keywords can be grouped into categories based on tasks as follows.

Tasks	Description
• Arrays	Defining, creating, and manipulating arrays of data
• Collection Object	Create/modify a collection object
• Compiler Directives	Automating compiler behavior
• Control Flow	Loop control and branching
• Conversion	Convert numbers and data types
• Data type	Setting and verifying data types
• Dates and Times	Date and time expressions
• Directories/Files	File characteristics and paths
• Error	Trapping errors
• Financial	Financial functions
• Input/Output	Reading/writing to files, printer object
• Math Functions	Trig functions, log, exponent
• Miscellaneous	Run other programs, send keys
• Operators	Comparisons, Boolean, modulus
• Registry	Access the registry
• String Manipulation	Parsing, formatting strings
• Variables/Constants	Declaring variables

Array Keyword Summary

The **Array** keyword returns a variant that contains an array. The syntax is Array(arglist), where arglist is a comma-delimited list of values. These values are subsequently assigned to the elements of the array contained within the variant. An example is vDirection=Array("N", "E", "S", "W"). vDirection(1) is equal to "E".

The **IsArray** keyword is used to determine if a variant variable is an array. A Boolean value is returned where a true indicates an array and a false indicates a non-array variant.

The **Option Base** statement is used at a module-declarations section level to declare the lower bound subscript for arrays. The Visual Basic default is "0". Accordingly, if you declare an array with a statement like Dim iCount(10) as integer. The First element of the iCount is iCount(0). The last element of iCount is iCount(10). Eleven integer elements are declared with a default option base statement.

Dim, **Private**, **Static**, and **Public** keywords are used for dimensioning arrays. Further information on these keywords is readily available with the use of Visual Basic Help.

The **Lbound** keyword is used to determine the smallest available subscript or the lower boundary for the indicated array. The required syntax is LBound(arrayname[, dimension]), and the value returned is a long. Dimension is optional for use with multidimensional arrays. In the example above, Lbound(iCount) is equal to zero.

The **Ubound** keyword is used to determine the largest available subscript or the upper boundary for the indicated array. The required syntax is UBound(arrayname[, dimension]) and the value returned is a long. Dimension is optional for use with multidimensional arrays. In the example above, Ubound(iCount) is equal to 10.

The **Erase** statement is used to reinitialize the elements of fixed-size arrays and de-allocate dynamic-array storage space.

A **dynamic array** is an array whose dimension can be changed during run time. The dynamic array can also be made multidimensional during run time. Declaring an array without any value creates a dynamic array. Dim vZone() is declared as a dynamic array of variants. The default data type in Visual Basic is the variant.

The **ReDim** statement is used to reallocate storage space for a dynamic array. The **Preserve** keyword is used in conjunction with ReDim, allowing the array to be redimensioned without losing any data. This is useful when an array needs to be increased in size to add a new element.

Collection Object Keyword Summary

A **Collection** object permits an ordered set of items to be referred to as a unit. Add, remove, and item methods are used to access the objects in the collection. Reference Microsoft Help for more information on this keyword.

Compiler Directive Keyword Summary

The **#Const Directive** and **#If...Then...#Else Directive** are used to conditionally compile code. Reference Microsoft Help for more information on this keyword.

Control Flow Keyword Summary

The **control flow** keywords are used for loops, logical decisions, accessing procedures, branching, and to pause or exit a program.

The **GoSub** keyword is used to create a subroutine within a procedure. The required syntax is GoSub *label.* Label marks an area of code with a name terminated with a colon (:). The location of the subroutine, as specified with *name*, must reside within the same procedure and terminate with a **Return**. An **Exit Sub** keyword is used to mark the beginning of the subroutine area in a procedure. The Exit Sub prevents the normal procedure flow from executing the subroutine code.

The **GoTo** statement causes an unconditional branch within a procedure. The required syntax is GoTo *label*. More structured flow controls are available for executing a branch.

The **On Error** statement is used for error-handling routines. You would use an error-handling statement like this if, for example, you were accessing a file from within your program. If the file didn't exist, or if the drive wasn't ready, your error-handling routine would deal with the problem. The error-handling statement is On

Error GoTo *label*, where label is the location of your error-handling routine within the current procedure. On Error Resume Next disregards the error and moves on to the next statement. On Error GoTo 0 disables all errors within the current procedure. To obtain a listing of error numbers and an associated error description, use Microsoft Help and search for "trappable errors."

The **On...GoSub** and **On...GoTo** statements are legacy Visual Basic commands that are still supported. Newer statements such as Select Case provide a more structured and efficient approach. Reference Microsoft Help for more information on these statements.

DoEvents() is used to transfer control to the operating system. When the operating system finishes processing the messages in queue, control is returned back to the program.

The **End** statement initiates an end to a procedure or a logic block. Typical End statements are as follows: End, End Function, End If, End Property, End Select, End Sub, End Type, End With. The End statement used by itself provides a way to force your program to halt.

The **Exit** statement is used to provide a means to exit a logic block. Exit Do, Exit For, Exit Function, Exit Property, and Exit Sub provide a clean method to exit the respective block of logic.

The **Stop** statement is similar to a breakpoint, in that program execution is halted and all the variables are maintained. A Stop can be placed anywhere in the code.

The **Do...Loop** statement is used to execute a block of code. Do While some condition is true and Do Until some condition is true are two methods of executing a Do...Loop. At any time within the loop, if an Exit Do is encountered, program flow will continue immediately after the Loop statement.

Loops can be contained within loops, which is commonly called nesting. The Loop Until or Loop While will always execute at least one pass of the Loop code. On the other hand, the Do Until or the Do While may go into the loop with the condition satisfied and immediately exit the loop. An example of Do...Loop syntax follows.

Form1

```
Do Until or Do While some condition is true
        Code
        If something Exit Do  (no limit to amount of Exit Do's)
        Code
Loop
```

Form2

```
Do
        Code
        If something Exit Do  (optional no limit to number of Exit Do's)
        Code
Loop Until or Loop While some condition is true
```

The **For...Next** statement is used to repeat a block of code for a specified number of times. A scratch-pad counter variable is required, as well as a starting count and an ending count. The default increment for the counter is one; however, the incremental value called a **Step** can be specified. The step value can be positive or negative. Additionally, the For...Next statements can be nested. Any time an Exit For is encountered, the program flow will continue immediately after the next statement. With a step value of -1, the start count should be greater than the end count. An example of For...Next statement syntax follows.

```
For variable = start to end Step (default 1 can be + or - any value)
        Code
        If something Exit For (optional no limit to the number of Exit For's)
        Code
Next variable
```

The **While...Wend** statement is another loop-based flow-control that executes the contained code as long as the given condition is true. The same functionality can be achieved with the Do...While loop. The Do...While loop statement is inherently more structured and flexible. Reference Microsoft Help for more information on this keyword.

The **With** statement executes a series of events on an object or a user-defined type. The following is an example of a With statement, as applied to a label object.

```
With Label1
 .AutoSize = True
 .BorderStyle = 1
 .Caption = “This was done with a With Statement”
End With
```

The **Choose** function returns a value from a list of arguments based upon a provided index. The required syntax is: variable = Choose(index, “choice1”, “choice2”, ... “choiceX”). If index = 1 then variable becomes equal to “choice1”.

The **If...Then...Else** statement is a very common decision-making statement. It can be presented in either line form or block form. The syntax of an If...Then...Else statement in line form is illustrated below. If more than one statement is required, use a colon (:) to separate them.

If some *condition* Then *statement: statement* Else *statement: statement*

The following is the required syntax of an If...Then...Else statement in block form. This is the preferred method, because the code is better structured and readable.

```
If some condition Then
Statements

    ElseIf some other condition Then {ElseIf is optional;can have more than 1}
    ElseIf Statements

Else: {optional}
ElseStatements
End If
```

The **Select Case** statement is a very powerful decision-making tool. The syntax for Select Case is as follows.

Select Case *value2beTested*

```
Case test1, test2,... testn
    statements
Case test1, test2,... testn
    statements

Case test1, test2,... testn
    statements
        •
        •
        •
Case test1, test2,... testn
    statements
Case Else
    Case else statements
End Select
```

The test expression for each case argument, depicted in the previous Select Case statement, can be single numeric values or strings. They can also be range expressions that use the **To** keyword, such as "1 to 9" or "a to f". In addition, logical operators such as **Is** > 6 or **Is** < 3 are permitted. Once a case clause is satisfied, the Select Case statement is ended and program flow continues after the End Select Statement.

The **Switch** function evaluates a list of functions and returns a variant value for an expression associated with the first expression in the list that is true. Reference Microsoft Help for more information on this function keyword.

The **IIf** function returns one of two possibilities, based on the evaluation of an expression. The syntax for this function is Iif(*expression, true value, false value*). The following example colors the background of a label control to green when the expression is true and to red when the expression is false. QBColor(10) paints the label background green. QBColor stands for Quick Basic Coloring scheme.

```
Dim hand As Boolean

hand = False
Label1.BackColor = IIf(hand, QBColor(10), QBColor(12))
```

Conversion Keyword Summary

The **Chr**(*character code*) function returns a string value for a specified character code. For example, Chr(10) returns a linefeed character. Chr$(13) is also acceptable syntax, and this example will return a carriage return to a string.

The **Asc**(*string*) function performs the complementary function of Chr. An integer code representing the character code is returned for the first character of the string. Asc("1") returns a value of 49.

The **Val**(*string*) function returns the first series of numbers found in a string. Characters such as spaces, tabs, and linefeeds are ignored. But the first non-number character other than those previously mentioned halts the conversion process. The plus sign and minus sign are recognized characters. The expression Val(" - 1 234 5678 9abcdefg") returns a value of -123456789. This function also recognizes the hexadecimal prefix &H and the octal prefix &O. Values with these prefixes are converted to decimal. The expression Val("&H1ABCD") returns a value of 109517.

The **Str**(*number*) returns a variant data type with a string representation for the supplied number. A leading space is always provided with any positive number conversion. If the number is negative, a minus sign is the leading character. Str(1.2345) returns " 1.2345". Str(-1234.5) returns "-1234.5".

The **Format** (*expression,"format"*) function operates on both numbers and strings. When the expression is a number, the result is identical to the string function except that a format argument is provided to structure the string. For example, you can specify the number of decimal places, the insertion of commas, the overall number of characters, trailing zeros, and so on. Table 1.04 illustrates some of the common format argument symbols and the associated functions. If the format expression is a string, the format arguments are typically provided to change all the characters of the string to either all upper case or all lower case characters.

In addition to the user-specified format structures using symbols, a number of "named format" structures are available. Formatting examples are given in the following list. For additional information, review the Microsoft Help files.

Symbol	Description
#	Digit placeholder only print spaces accordingly
.	Decimal placeholder
,	Thousands separator
- + $ (space) X	Literals are printed exactly as specified
<	Operates on strings to change all characters to lower case
>	Operates on strings to change all characters to upper case

Table 1.04

```
Format( 1234.50, "00000.00") => 01234.50
Format( 1234.50, "#####.##") => 1234.5
Format( 1234.50, "####0.00") => 1234.50
Format( 1234.50, "$###0.00") => $1234.50
Format( 1234.50, "$#,##0.00") => $1,234.50
Format( 1234.50, "$#A##0.00") => $1A234.50 ' any literal will be transferred
Format(  Now, "Medium Date") => 29-Oct-00
Format(  Now, "mmmm-yy") => October-00
Format("KaitLyn", "<") => kaitlyn 'convert all to lower case
Format("Kallie", ">") => KALLIE  'convert all to upper case
```

The **LCase**(*string*) function returns a string that is all lower case. The **UCase**(*string*) function returns a string that is all upper case. The **Hex**(*number*) function returns a string that is hexadecimal. The **Oct**(*number*) function returns a string that is Octal. The **Int**(*number*) and **Fix**(*number*) functions return the integer value of a number. The only difference between these two is the way each one handles a negative number. Given a number of –9.4, while the result of Int(-9.1) is a –10. The result of Fix(-9.1) is –9.

String Manipulation Keyword Summary

The **StrConv**(string, conversion, LCID) function returns a string converted as specified in the conversion argument. "LCID" is an optional location identifier. The statement StrConv("cape hatteras", vbProperCase) returns the string Cape Hatteras. Some of the possible conversion arguments are the Visual Basic constants listed in Table 1.05.

vbUpperCase	Convert to upper case characters.
vbLowerCase	Convert to lower case characters.
vbProperCase	Convert to upper case the first character of each word.
vbUnicode	Convert to 16-bit (2-byte) Unicode characters.
vbFromUnicode	Convert 2-byte Unicode to the Ansi character set.

Table 1.05

A standard universal character set controlled by International Standard Organization is called **Unicode**. Unicode is a 16-bit character set that supports 65,536 characters. The extended character set includes both graphics and foreign language characters.

The **Len**(*string or variable*) function returns a long containing the string length. Len("1234") returns a value of 4. Dim A as integer. Len(A) returns a value of 2. An integer data type consists of 2 bytes.

The **String**(*number," string character"*) function creates a string or variant with the number and string character specified. The statement String(5,"^") returns ^^^^^. The **Space**(*number*) function creates a string or variant with the specified amount of spaces. Space(20) returns a string containing 20 space characters.

The **Lset** and **Rset** functions are used to left and right justify string expressions, respectively. The syntax for these functions is Lset *string of fixed specified length* = *string to be inserted.* As an example, examine the following code.

```
MyStr = Space(10)
MyStr1 = "1"
LSet MyStr = MyStr1
```

The result of the Lset is MyStr consisting of the character 1 followed by nine spaces. If you change Lset to an Rset, the result is nine spaces followed by the character 1.

The **LTrim**(*string*) function removes any leading spaces from a string. The **RTrim**(*string*) function removes any trailing spaces from a string. The **Trim**(*string*) function removes both leading and trailing spaces from a string.

The **Left**(*string, length*) function returns the specified length of character from the left side of the specified string. Left("12345", 2) returns a string consisting of "12".

The **Right**(*string, length*) function returns the specified length of character from the right side of the specified string. Right ("12345", 2) returns a string consisting of "45".

The **Mid**(*string, start, length*) function returns a string of specified length that begins at the start position indicated. If the length argument is omitted, all the characters from the start position to the end of the string are returned. Mid("12345", 3, 1) returns a string containing the character "3". Mid("12345", 3) returns a string containing the characters "345".

The **InStr**(*optional start position, string, string expression sought, optional compare*) function returns the first occurrence of the string expression sought within the specified string. This function searches for the string going left to right. The method of comparison is also programmable. A binary compare checks for an exact character match and is specified with a vbBinaryCompare constant. A textual compare is not case sensitive and is specified with a vbTextCompare constant. A vbDatabaseCompare type is also available, which works with a Microsoft Access Database. Several examples are provided below.

```
Position = InStr(1, "abcABCabcABC", "a", vbTextCompare)  => 1
Position = InStr(2, "abcABCabcABC", "a", vbTextCompare)  => 4
Position = InStr(2, "abcABCabcABC", "a", vbBinaryCompare) => 7
Position = InStr(5, "abcABCabcABC", "A", vbBinaryCompare) =>10
```

The **InStrRev**(*string, string expression sought, optional start position, optional compare*) function is similar to the InStr function, except the syntax-argument arrangement differs slightly and the operation is executed from the end of the string to the beginning of the string.

```
Position = InStrRev("abcABCabcABC", "A", 5, vbBinaryCompare)  => 4
Position = InStrRev("abcABCabcABC", "A", 12, vbBinaryCompare) =>10
Position = InStrRev("abcABCabcABC", "ABC", 10, vbBinaryCompare) =>4
```

The **StrComp**(*string1, string2, optional compare*) function returns a variant based on the string comparison. The optional compare parameter can be binary, textual, or database-related. The result yields a zero for a match, -1 if string 1 is less than string 2, and +1 if string 1 is greater than string 2.

Variables/Constants Keyword Summary

The **Dim** statement is used to declare the type of a variable and to allocate the appropriate amount of bytes. Variables declared at the procedure level with the Dim statement are only available within the procedure. Variables declared at the module level with the Dim statement are available to all the procedures contained in the module. When variables are initialized, a numeric variable is initialized to 0, a variable-length string is initialized to a zero-length string (""), and a fixed-length string is filled with zeros. Variant variables are initialized to an empty. Each element of a user-defined type variable is initialized as if it were a separate variable.

The **Private** statement is used to declare a variable at the module level as being private. Variables declared using the **Public** statement are available to all procedures in all modules in all applications. Public variables are declared at the module level.

The **Static** statement is used at the procedural level to declare a variable type and to allocate the appropriate amount of bytes. The contents of the Static-declared variable, however, are preserved as long as the program is running.

Constants make the code much more readable. Constants declared in a Sub, Function, or Property procedure are local to that procedure. A constant declared outside a procedure is defined throughout the module in which it is declared. When declared at the module level as Public, a constant is then available to all modules and procedures within a program. The required syntax is [Public or Private] Const constname [As type] = expression. Public and Private are optional. Visual Basic provides a number of intrinsic constants pertaining to the system and installed controls. These intrinsic constants can be explored with the Object Browser.

The **Option Explicit** statement, if used, occurs at the module level before any procedures. Option Explicit requires that all variables be declared with either a Dim, Private, Public, Static, or ReDim statement.

Deftype statements allow you to select an alphabetic range as an automatic data type. For example, DefStr A-E: B="abcd". Any variable name that starts with a character from A to E is implicitly declared to be a string. A variable starting with any of the letters within the range can always be explicitly declared a different data type. All data types, except decimal, are supported.

A variant data type can contain any data type. Functions are provided to determine the data type contained in a variant. These functions are **IsArray()**, **IsDate()**, **IsEmpty()**, **IsNumeric()**, **IsObject()**, **IsNull()**, and **IsMissing()**. In addition, the function VarType(variable name) returns an integer for the data type contained in a variant. Intrinsic Visual Basic constants are available to identify each data type, as illustrated in Table 1.06.

The **Me** keyword is used to reference the form whose code is currently running. Another way to specify the current form in code is to use the Me keyword.

VB Constant	*Value*	*Description*
vbArray	8192	Array
vbBoolean	11	Boolean value
vbByte	17	Byte value
vbCurrency	6	Currency value
vbDataObject	13	A data access object
vbDate	7	Date value
vbDecimal	14	Decimal value
vbDouble	5	Double-precision floating-point number
vbEmpty	0	Empty(not intialized)
vbError	10	Error value
vbInteger	2	Integer
vbLong	3	Long integer
vbNull	1	Null (no valid data)
vbObject	9	Object
vbSingle	4	Single-precision floating-point number
vbString	8	String
vbUserDefinedType	36	Variants that contain user-defined types
vbVariant	12	Variant (used only with array of variants)

Table 1.06

Operators Keyword Summary

^ Operator is an arithmetic operator used to raise a number or a value to the power of an exponent. For example, Number = 2 ^ 4 is equal to 16.

- Operator is an arithmetic operator used for subtraction.

* Operator is an arithmetic operator used for multiplication.

/ Operator (Forward Slash) is an arithmetic operator used for division. The result is a floating-point number.

\ Operator (Back Slash) is an arithmetic operator that performs division and yields an integer result.

The **Mod** Operator performs a division operation where the result is the remainder.

The **+** Operator performs an addition operation on two expressions. The + Operator can also be used to add strings; however, it's recommended that the & Operator for concatenation of string data types be used to eliminate any ambiguity that may result.

The **&** Operator performs a string concatenation of two expressions. If the expression is not a string, it is automatically converted to a string variant for the operation.

The = Operator is used to equate a value to a variable or an object property.

Available Comparison Operators are listed in Table 1.07.

Operator	*Description*
=	Equal
<>	Not equal
<	Less than
>	Greater than
<=	Less than or equal to
>=	Greater than or equal to

Table 1.07

The **Like** Operator is used to compare two expressions based upon the setting of the Option Compare statement. The default option compare property is Option Compare Binary, which is case sensitive.

The **Is** Operator can be used to compare two object references. The expression is evaluated as true if both object1 and object2 refer to the same object.

The **Not** Operator is used to cause a logical inversion. The line of code Dim vData :vData = Hex((Not 1 And 15)) returns a result of E hexadecimal. In binary, the number 1 is the nibble 0001. The inverse of the nibble is 1110, which equates to E in hexadecimal.

Logical And Function		
Exp1	Exp2	Result
0	0	0
0	1	0
1	0	0
1	1	1

Table 1.08

The **And** Operator requires that both expressions be true for the result to be true. The required syntax is result = expression1 And expression2.

An **And** function was used in the Not example to select only the last nibble or 4 bits of the result. The result of 3 And 7 = 3. The logic Table 1.08 illustrates an And function. The logic Table 1.08 illustrates an And function.

Logical OR Function		
Exp1	Exp2	Result
0	0	0
0	1	1
1	0	1
1	1	1

Table 1.09

The **Or** Operator requires that any expression be true for the result to be true. The required syntax is result = expression1 Or expression2. The logic Table 1.09 illustrates an Or function. The result of 3 Or 7 = 7.

The **Xor** (Exclusive Or) Operator requires that one expression be true, but only one for the result to be true. The required syntax is result = expression1 Xor expression2. The logic Table 1.10 illustrates an Xor function. The result of 3 Xor 7 = 4.

Logical XOR Function		
Exp1	Exp2	Result
0	0	0
0	1	1
1	0	1
1	1	0

Table 1.10

The **Eqv** (Equivalence) Operator is an Exclusive Nor or Coincidence function. If both expressions are false or both expressions are true, the result is true. The required syntax is result = expression1 Eqv expression2. The logic Table 1.11 illustrates an Eqv function. The result of 3 Eqv 7 = -5.

Logical Eqv Function		
Exp1	**Exp2**	**Result**
0	0	1
0	1	0
1	0	0
1	1	1

Table 1.11

The **Imp** (implication) Operator is used to perform a logical implication function. The required syntax is result = expression1 Imp expression2. The logic Table 1.12 illustrates an Imp function. The result of 3 Imp 7 = -1.

Logical Imp Function		
Exp1	**Exp2**	**Result**
0	0	1
0	1	1
1	0	0
1	1	1

Table 1.12

Dates & Times Keyword Summary

The **Now** function returns a variant (Date data type) that contains the current date and time according to your computer's system date and time preference. The line of code Dim vNow: vNow = Now() returns "1/13/00 4:09:09 PM".

The **Date** function returns variant containing the current date. Dim vDate: vDate = Date returns "2/13/00".

The **Time** function returns variant containing the current time. Dim vTime: vTime = Time returns "4:19:25 PM".

The **Hour** function returns a variant containing the hour of time. The argument can be a variant, a string, or a numeric. The result ranges from 0 to 24. An example of the hour function is Dim vHour: vHour = Hour(Now()) =>"22".

The **Minute** function returns a variant containing the minute of the day. The argument can be a variant, a string, or a numeric. The result ranges from 0 to 59. An example of the minute function is Dim vMinute: vMinute = Minute(Now()) => "29".

The **Second** function returns a variant containing the second of the day. The argument can be a variant, a string, or a numeric. The result ranges from 0 to 59. An example of the second function is Dim vSecond: vSecond = Second(Time)=> "15".

The **DateSerial** function returns a variant containing a date data type for a specified year, month, and day. The syntax is DateSerial(year, month, day).

The **DateValue** function returns a variant containing a date data type for a specified year, month, and day. The syntax is DateValue(" date string"). Dim vDate: vDate = DateValue("oct 29 2000"). VDate contains 10/29/00.

The **TimeSerial** function returns a variant with a date data type for a specified time. The syntax is TimeSerial(hour, minute, second).

The **TimeValue** function returns a variant of date data type containing time. The syntax is TimeValue("string time"). Dim vTime: vTime = TimeValue("10:59") returns 10:59:00 AM.

The **DateAdd** function permits you to determine a date based upon a specified interval of time added to a given date. The returned value is a date variant. The required syntax is DateAdd(*interval*, *number*, *date*). The interval-type argument specifies whether to add weeks, hours, days, or weekdays (since Visual Basic allows you to add X days or X Wednesdays, for example, to the date). The number argument is the amount to add and the date argument is the starting date. Use the Help files to get additional information.

The **DateDiff** function is similar to the DateAdd function, except now the result is an interval of time in the specified units between date1 and date2. Use the Help files to get additional information.

The **DatePart** function allows you determine where a specified date occurs relative to a specified interval unit. For example, what week of the year contains the day 22 Oct 00 ? Use the Help files to get additional information. The Date statement will set the current system date. The Time statement will set the current system time.

Input & Output Keyword Summary

The **FileCopy** statement will copy a specified file to a specific location. The required syntax is FileCopy source, destination. See the following sample code.

```
Dim source, destination
source = "c:\test\test.txt"
destination = "c:\test\test1\test.txt"
FileCopy source, destination
```

The **FileDateTime** function returns a variant string consisting of a specified file's last modified date. The syntax is FileDateTime(filename string). Fdt = FileDateTime("c:\test\test1\test.txt").

The **FileLen** function returns a long integer with the specified file's length in bytes. The required syntax is FileLen(filename string).

The **FileAttr** function returns a long, indicating file mode for an open file. The required syntax for all 32-bit systems is FileAttr(filenumber, 1).

The **GetAttr** function retrieves the attribute information of a specified file or folder. The required syntax is GetAttr(pathname). According to Table 1.13, an integer value is returned. It's possible for a file to have more than one attribute set. Boolean operations are used to determine the individual bits set.

The **SetAttr** statement is used to set the attributes of a given file. The file must be closed to perform this task. The required syntax is SetAttr pathname, attributes.

The **Open** statement opens a disk file for input or output. The required syntax is Open pathname For mode [Access access] [lock] As [#]filenumber [Len=reclength]. Pathname is a string expression including drive, any folders, and the filename. The file mode may be input, output, append, binary, or random. Optional keywords specifying the access operations permitted on the open file are read, write, or read write. The Lock option keyword specifies the operations restricted on the open file by other processes: Shared, Lock Read, Lock Write, and Lock Read Write. The file number is

Vbconstant	*Value*	*Description*
vbNormal	0	Normal.
vbReadOnly	1	Read-only.
vbHidden	2	Hidden.
vbSystem	4	System file.
vbDirectory	16	Directory or folder.
vbArchive	32	File has changed since last backup.

Table 1.13

required, and is valued in the range from 1 to 511, inclusive. The FreeFile function will provide you with the next available file number. The record length parameter for files opened for random access is the record length. For sequential files, the record length parameter is the number of characters buffered. Later chapters of this book will be using file access for storing and retrieving information.

The **FreeFile** function returns an integer value for the next available file number. The Close statement terminates the input/output operations on an opened file. The required syntax is Close(filenumber). If (filenumber) is omitted, all active open files are closed. All buffer space associated with the closed file is released. Sometimes, due to the way the Windows operating system buffering process works, this command is not satisfied as quickly as possible. The following Reset command assigns a priority to the close operation.

Another statement called **Reset** will close all active open files and place all the buffer information to the disk. The Reset command forces the operating system to respond faster when closing files.

The **EOF** function returns a Boolean result of True when the end of a file opened for Random or sequential Input has been reached. The required syntax is EOF(filenumber).

The **Print #** statement writes data to a open sequential file. The required syntax is Print #filenumber, data. The file number is typically determined with the FreeFile statement and is the same as the file open number. Data written with Print # is usually read from a file with Line Input # or Input statements

The **Line Input #** statement is used to retrieve a line of data from an opened sequential file. The required syntax is Line Input #filenumber variable name. The returned value is a string.

The **Write #** statement also writes automatically comma-delimited data to a sequential open file. Unlike the Print # statement, the Write # statement inserts commas between items and quotation marks around strings as they are written to the file. The required syntax is Write #filenumber, data.

The **Input #** statement reads data from an open sequential file. The required syntax is Input #filenumber, variable name.

The **Width #** statement is used to set the output line width for an opened file. The required syntax is Width #filenumber, width. The width value determines how many characters appear on a line before a new line is started.

The **Put** statement is another method of writing data to an open file. The required syntax is Put #filenumber, record number, data.

The **Get** statement is generally used to retrieve data from a file processed with a Put statement. The required syntax is Get #filenumber, record number, variable name.

The **Seek** statement is used to provide positioning information for records when using a Put or Get statement.

Directories & Files Keyword Summary

The **ChDrive** statement changes the current drive setting. The syntax is Chdrive drive. An example is Chdrive "A".

The **Dir** function returns a string representing the name of a file, directory, or folder that matches a specified pattern or file attribute, or the volume label of a drive. The syntax is string = Dir [(pathname[, attributes])]. As an example, String = Dir("c:\", vbVolume) returns the volume for the path specified.

The **ChDir** statement changes the folder or directory for the current drive. This statement does not change the current drive. The statement ChDir "D:\WINDOWS\command" changes the default folder on drive D.

The **CurDir** function returns a variant string with the current path setting of the specified drive. String = CurDir("D") returns the current path for drive D.

The **MkDir** statement creates a new folder in the specified or current path. The syntax is MkDir "path". The statement MkDir "A:\TL" creates a folder called "TL" on A drive.

The **RmDir** statement removes the folder in the specified or current path. The syntax is RmDir "path". The statement RmDir "A:\TL" removes a folder called "TL" on A drive.

The **Name** statement renames a drive file. The syntax is Name old name As new name. The Name statement renames a file and, if necessary, moves it to a different directory or folder. The statement Name " a:\ vb.txt" As "a:\tl\vb.txt" moves a file called "vb.txt" located in the root of A drive to folder in A drive called TL.

Miscellaneous Keyword Summary

The **Beep** statement causes the PC computer to emit a beep sound. The sound level, frequency, and duration are determined by the computer hardware.

The **Shell** function runs an executable program and returns a variant representing the program's task ID. The required syntax is Shell(pathname,optional windowstyle). You can use shell to run any executable program, including batch files, from a Visual Basic window. The optional window styles are hidden, normal with focus, minimized with focus, maximized with focus, normal without focus, and minimized without focus.

The **AppActivate** statement does not launch an application; it changes the focus to another project that is running. The syntax is AppActivate title[, wait]. Once AppActivate has activated a window, you can send keystrokes to the active window using the SendKeys statement.

The **SendKeys** statement sends one or more keystrokes to the active window. This feature is useful for running demonstrations and such. The syntax is SendKeys string[, wait].

The **CreateObject** function is used to create and return a reference to an ActiveX object. The required syntax is CreateObject(*class,[server name]*). Class is string, which references a class object. The server-name argument is optional and references the server name where the object will be created. The following provides an example with Microsoft Excel. An object containing an Excel sheet is created and made visible. Text is then sent to cell "D4". The application is then closed and the object released.

```
Dim Excelx As Object
Set Excelx = CreateObject("Excel.Sheet")
Excelx.Application.Visible = True
Excelx.Application.Cells(4, 4).Value = "VB accessing cell D4"
Excelx.SaveAs "C:\example.XLS"
Excelx.Application.Quit
Set Excelx = Nothing ' release the object Excel
```

The **GetObject** function is used to reference an ActiveX object. In the following example, the text string, which was inserted into "C:\example.xls" cell "D4" with a CreateObject function, is retrieved in variable vCell using a GetObject function.

```
Dim Excelx As Object
Set Excelx = GetObject("C:\example.XLS")
Excelx.Application.Visible = True
Excelx.Parent.Windows(1).Visible = True
Dim vCell
vCell = Excelx.Application.Cells(4, 4).Value
Excelx.Application.Quit
Set Excelx = Nothing
```

The **QBColor** function is used to change the color property of various objects. The required syntax is QBColor(*color*). The statement Label1.BackColor = QBColor(10) sets the back color of label1 to green. This function dates back to an early version of Visual Basic called "Quick Basic." There are 16 possible colors, presented in Table 1.14.

The **RGB** function is another method of changing color-related properties of objects. This function provides the capability to adjust each of the red-green-blue colors from a value of 0 to 255. The required syntax is RGB(red, green, blue). Table 1.15 illustrates several colors and the associated RGB values.

Conclusion

The material covered so far satisfies the aim of this chapter to provide a whirlwind overview of the Visual Basic environment. The Keyword categories examined thus far are within the scope of the software covered in this book. Other Keyword categories are left for you to examine using Microsoft's Help files. Control objects will be discussed in later chapters as various projects are developed.

Number	Color	Number	Color
0	Black	8	Gray
1	Blue	9	Light Blue
2	Green	10	Light Green
3	Cyan	11	Light Cyan
4	Red	12	Light Red
5	Magenta	13	Light Magenta
6	Yellow	14	Light Yellow
7	White	15	Bright White

Table 1.14

Color	Red Value	Green Value	Blue Value
Black	0	0	0
Blue	0	0	255
Green	0	255	0
Cyan	0	255	255
Red	255	0	0
Magenta	255	0	255
Yellow	255	255	0
White	255	255	255

Table 1.15

Chapter 2
SERIAL COMMUNICATIONS BASICS

INTRODUCTION

For some reason, the term "serial communications" always implies some kind of high-tech, complicated communications procedure. In fact, serial communications is simply the basis for common everyday communications. When you talk, you emit vocal sounds called phonemes, which another person receives and mentally assimilates into words. When you read this book, your eyes scan from left to right, gathering the letters delimited by spaces into words. Likewise, serial communications is the basis for communication among computers.

Human vocal communication is more complex than computer communication by an order of magnitude. Computers communicate with a fixed amount of bits, at a fixed rate, with a fixed framing structure. Human vocal communications, on the other hand, is extremely dynamic.

IBM-based personal computers are equipped with quite an assortment of serial ports. The keyboard port, PS2 mouse port, RS-232 serial port, parallel port, USB port, network port, video port, and modem port are all serial communication ports. This chapter focuses on the RS-232 standard of serial communication.

There are two basic forms of serial communications, synchronous and asynchronous. With **synchronous communications**, the data is transmitted in sync with a clock signal. **Asynchronous communications** eliminates the clock and adds extra bits to indicate the start and the end of the data. RS-232 ports provided with a computer support asynchronous serial communications.

The configuration of a serial character is sometimes referred to as a **frame**—an appropriate term, since all the character's communication bits are contained in it. The bits contained within the character frame are **Start bit, Data bits**, **Parity bit**, and **Stop bit**.

Most computers utilize an integrated circuit called a **UART** to handle asynchronous serial communications. UART is an acronym meaning Universal Asynchronous Receiver Transmitter. This chip typically operates at five-volt logic levels and takes care of all of the serial-to-parallel and parallel-to-serial conversion for the CPU. Both the receiver input and transmitter output are routed to an RS-232 driver chip to interface with the outside world. If you look at the asynchronous line at the UART, the idle state sits at five-volts or at a logic one. When a character is transmitted, the line goes low for one bit constituting the Start bit. This falling edge transition alerts the receiver that a character is traveling down the line.

Once the Start bit is launched, the Data bytes follow. The number of bits that make up a character can be specified as 5, 6, 7, or 8. Most serial devices transmit data using either 7 or 8 data bits. Seven Data bits are all the bits required to transmit true ASCII characters. **ASCII** is a character set that contains alphanumeric, formatting, and control characters similar to the characters found on standard keyboard keys. ASCII consists of 2^7, or 128 different characters. An ASCII table is included in the appendix. Eight Data bits are required to transmit bytes of information or software code.

An optional bit called the Parity bit may be transmitted along with the data to provide a small amount of error detection. Valid parity values are Even, Mark, None, Odd, and Space. **Mark Parity** means that a logic one called a mark in telemetry circles is transmitted as the Parity bit. Likewise, **Space Parity** means that a logic zero called a space is transmitted as the Parity bit. **None Parity** indicates that no Parity bit of any sort is transmitted. With **Even Parity**, the number of ones in the data byte and the Parity bit is always an even number. For **Odd Parity**, the number of ones in the data byte and the Parity bit is always odd. In all cases, the Parity bit is set accordingly to ensure the specified parity. The best choice for the Parity bit is none. Mark and Space Parity choices are seldom used, since they just add time on to the transmission time of the character. Given the minimum error detection offered by Even Parity and Odd Parity, the decrease in transmission time gained by the elimination of the Parity bit is the better bargain. Parity check will detect a single bit error in a character, but if an even number of bits is in error, it will be undetected. More robust error detection schemes will be investigated in later chapters.

The Stop bit is a logic one or a mark that ends the character transmission. Stop bits can be specified to be 1, 1.5, or 2 bits in length.

The **Baud rate** determines the electrical period of time required for each transmitted bit. This unit of measurement is named after Jean-Maurice-Emile Baudot. Baudot originated the idea of a 5-bit data character, which surpassed Morse code as the most commonly used telegraphic alphabet in the early 1900s. Quite frequently, Baud rate is erroneously used to indicate bits per second (bps). In the case of a hard-wired serial port to serial port connection, the Baud rate is identical to the rate of transfer of information (bps). There is 1 bit of information per Baud. Modern day modems, on the other hand, use sophisticated modulation schemes and compression to substantially increase the rate of information transfer. Each Baud can carry several information bits. Despite the hardware limitations of a voice-grade telephone line to somewhere below 4000 Hertz, 56K bit per second modems are common. The standard Baud rates supported by Visual Basic are 110, 300, 600, 1200, 2400, 9600, 14400, 19200, and 28800.

Serial communication data format is frequently expressed as Baud rate, Parity, Data bits, and Stop bit parameters. For example, 9600, N, 8, 1 indicates a Baud rate of 9600, no Parity, 8 Data bits, and 1 Stop bit. The framing of this type of data byte is illustrated in Figure 2.01. The whole character requires 10 bits. Therefore, at a Baud rate of 9600 with this data structure, the character throughput is roughly 960 characters per second. Figure 2.01 shows the transmittal of a byte equal to 01 hexadecimal.

An Electrical Industries Association (EIA) driver conditions this bit stream before interfacing with real-world devices. This driver adheres to a standard as defined by the EIA. Some of the popular standards are RS-232, RS-422, and RS-485. All personal computers come equipped with RS-232C serial ports. **RS-232C** stands

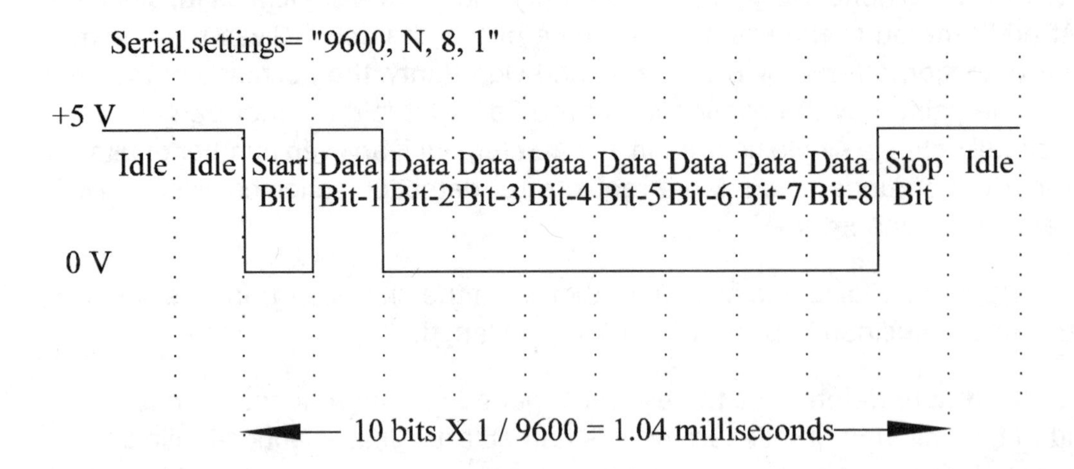

Figure 2.01

for Recommended Standard 232, and the suffix C indicates the latest revision. The formal EIA title for RS-232C is "Interface between Data Terminal Equipment and Data Communications Equipment Employing Serial Binary Data Interchange" and is dated August 1969. The RS-232 standard defines the mechanical connection characteristics, the functional description of each line, and the electrical signaling characteristic.

Serial Pin Out Description

The RS-232 standard specifies a 25-pin D-type connector, or DB-25 as the standard mechanical connector. Twenty-two of the 25 pins are defined. A male D-connector is specified to terminate on the computer equipment referenced as "DTE" by the standard. DTE means Data Terminal Equipment and always means the computer. As you would expect, the connecting equipment referred to as "DCE" by the standard terminates with a female D-type connector. Personal computer communications requirements reduce the 22 pins of the standard down to 10 pins. These pins and associated functions for a 25-pin D-type connector are illustrated below.

DTE(computer) 25-pin male

1 Protective ground
2 TD Transmit Data (output)

3 RD Receive Data	(input)
4 RTS Request To Send	(output)
5 CTS Clear To Send	(input)
6 DSR Data Set Ready	(input)
7 Common	(power supply common)
8 CD Carrier Detect	(input)
20 DTR Data Terminal Ready	(output)
22 RI Ring Indicator	(input)

Most personal computers today have deviated from the DB-25 mechanical standard of RS-232. In lieu of the DB-25 male, a DB-9 male is typically supplied. There was a time in the early days of the personal computer when the computer had both a standard DB-25 (typically used for an external modem connection) and a DB-9 (used to accommodate devices such as the mouse). The pins and associated functions for a 9-pin D-type connector are illustrated below.

DTE(computer) 9-pin male

1 CD Carrier Detect	(input)
2 RD Receive Data	(input)
3 TD Transmit Data	(output)
4 DTR Data Terminal Ready	(output)
5 Common	(power supply common)
6 DSR Data Set Ready	(input)
7 RTS Request To Send	(output)
8 CTS Clear To Send	(input)
9 RI Ring Indicator	(input)

The DTE functional description of these pins is as follows.

Carrier Detect (pin 1) A modem when a carrier frequency is detected sets an output. This input monitors the modem's carrier detect output.

RD Received Data (pin 2) Data is received from the TD of a remote device through this line. When connected to a remote device, RD will sit in a mark state derived from the TD output of the connecting device while communications is idle.

TD Transmit Data (pin 3) The data is transmitted out of this pin into the receive pin of the connecting equipment. The TD line is in a mark condition while the communications is in an idle state.

DTR Data Terminal Ready (pin 4) DTR is an output that connects to the connecting device's DSR input. The original intent of this complementary pair of data-control lines was some form of special hardware data-handshaking feature for data-flow control. Some devices use these data-control lines to indicate that a cable exists between the two devices.

Signal Common (pin 5) This line brings the interconnecting devices to the same point of reference for zero volts.

DSR Data Set Ready (pin 6) Data Set Ready is a data-control input that is connected to the DTR of the connected device. DSR is the companion control line to DTR.

RTS Request To Send (pin 7) Request to Send is a data-flow control output that connects to the connecting device's Clear to Send (CTS) input. Typically, RTS and CTS are used to implement a hardware data-flow control. If used for flow control, the computer (DTE) places the RTS in the mark state, indicating to the remote device that it's ready to receive data. If the receive buffer of the DTE should become full, a space is asserted on RTS instructing the interconnecting device to temporarily stop sending data.

CTS Clear To Send (pin 8) Clear to Send is a data-flow control input that is wired to the RTS of the interconnecting device. CTS is a data-flow control companion to RTS. If a mark is asserted on CTS, the DTE will continue to transmit data. Data transmission will be temporarily halted if a space is detected on CTS.

RI Ring Indicator (pin 9) The Ring Indicator is an input that monitors a ring-signal output from a modem. This output tracks with respect to the ring signal on the telephone line.

The term **handshaking** is used to indicate data-flow control. Data-flow control is critical for any slow responding data-transfer application. One example is a serial-based printer. Although printers do have data-collection buffers, the print job is frequently large enough that the print buffer is quickly filled. Data-flow control acts like a traffic cop, preventing the print buffer from overspilling and distorting the printout. The electrical characteristics of RS-232 specify a zero-crossing bipolar signal. A voltage of less than –3 volts represents a mark (logic 1) on a data line. A voltage of +3 volts or more indicates a space (logic 0) on a data line. A control line represents an off with a voltage less than –3 volts and an on with a voltage greater than + 3 volts. The maximum voltages are + 15 and –15, respectively.

A throwback to the old Teletype days, the negative voltage requirement of RS-232 does appear a bit odd today. Other newer standards such as RS-422 operate electrically from 0 volts to 5 volts with differential (complementary) outputs. Here, as one output is being driven high, the other output is being driven low. Although it takes two physical outputs and thus two wires for one data bit stream, these outputs are capable of achieving some very high Baud rates over some very long distances. At 9600 Baud distances, more than 10,000 feet are possible.

The RS-232C standard also imposes a limit on cable length to approximately 50 feet. For the most part, this limitation can be ignored if a high-quality shielded cable is used. Also, the RS-232 driver chips available today have excellent line-driving capabilities. It's not unusual to communicate several hundred feet at 9600 Baud with a good-quality shielded cable. The maximum transmission length is a function of the operating Baud rate. A greater distance of communications can be achieved at 110 Baud, compared to the distance obtained with a Baud rate of 28800.

Historically, the biggest problem with RS-232C communication has been the interconnection between devices—due to the wide assortment of adapters (such as DB9 to DB25), gender changers, and cable configurations. When dealing with an unknown configuration, the best approach is to confirm with a digital multimeter the voltage levels on pins 2 and 3 with respect to the common. The common is pin 5 on the DB-9 connector and pin 7 on the DB-25 connector. The pin that swings to a negative voltage is definitely the Transmit Data output. A typical input, when looked at with a digital multimeter, will float somewhere around 0 volts.

Serial Interconnection

The simplest connection to a write-only device, such as an Alphanumeric Display, requires only two wires (as illustrated in Figure 2.02). You could also have a serial device measuring some physical parameters configured to output a serial update

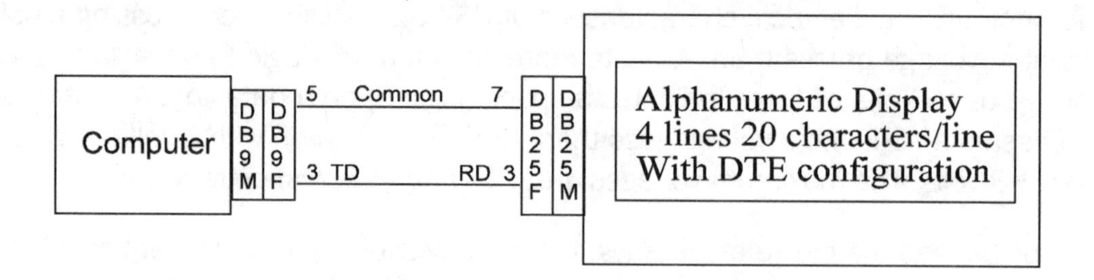

Figure 2.02

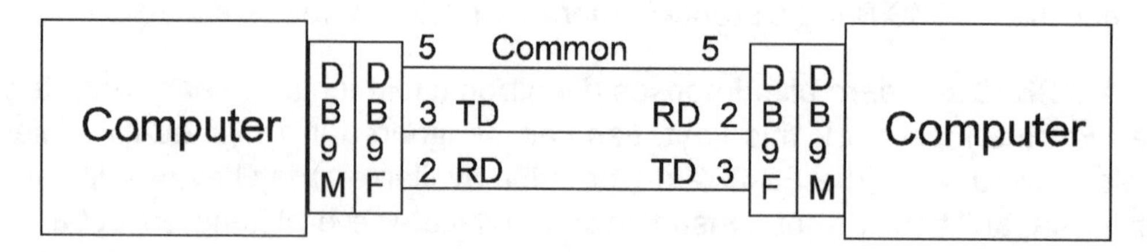

Figure 2.03

at a periodical rate. In this case, the computer becomes the receiver of the information, and again, only two wires are required for the connection.

The most common interconnection is the 3-wire, as illustrated in Figure 2.03. If handshaking is required today, it's typically accomplished by software through XON/XOFF Data-Flow Control Protocol. With XON/XOFF protocol, the receiver sends an XOFF character instructing the transmitter to pause the transmission and an XON character signifying that the transmission is to resume. Typically, the XON character is ASCII DC1 or 11 Hex. The typical XOFF character is ASCII DC3 or 13 Hex. An ASCII table can be found in this book's Appendix.

Figure 2.04 illustrates the typical interconnection to a serial device with hardware handshaking. Another popular 3-wire configuration is illustrated in Figure 2.05. With this configuration, the interfacing device unnecessarily requires hardware handshaking, and feeding back the control signal to itself satisfies this requirement.

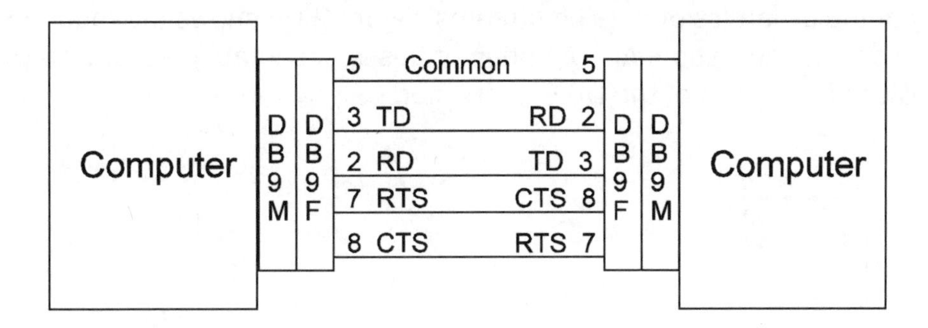

Figure 2.04

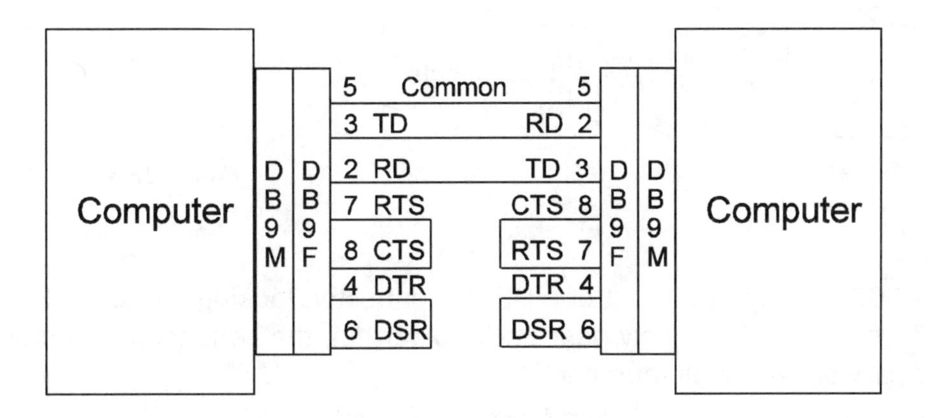

Figure 2.05

First Visual Basic Serial Port Examination Program

Now you'll use Visual Basic to verify and reinforce the information presented above. Your first Visual Basic program is going to introduce you to Microsoft's MSComm component, a little animation programming, and to the electrical characteristic of one of your computer's RS-232 serial ports. For this first program, you will need a 9-conductor cable with a DB-9 female connector at one end and a DB-9 male connector on the other end. A stand-alone DB-9 female connector that you can solder up with test wires is also required, as is a multimeter. An oscilloscope would be ideal. Oscilloscopes show voltage with respect to time, which is perfect for looking at the dynamics of data bit-streams. Figure 2.06 shows the required components and configuration. Any variation of this is acceptable, as long as you are able to access the pins of the DB-9 connector.

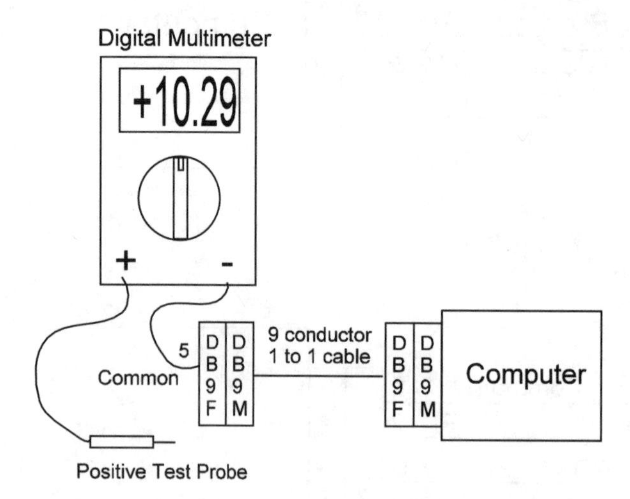

Figure 2.06

Figure 2.07 shows the pin out of a DB-9F connector, looking at the back (solder side). If you look very closely near each solder pin, the connector manufacturer will typically place the pin number.

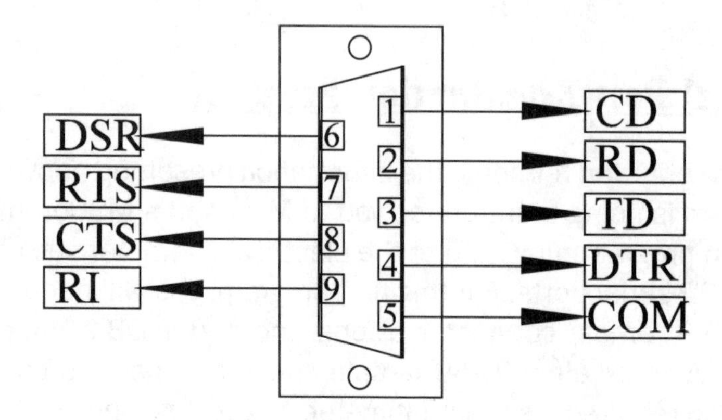

Figure 2.07

First Program

The software objective of this program is:

1) Mouse click on/off the control outputs DTR and RTS.
2) Show the status of control inputs CD, DSR, CTS, and RI.
3) Transmit a byte at a periodic rate.

Launch Visual Basic 6. Select new Standard Exe as the project type. Change the caption for Form1 to read "Serial 1." The following code is designed for a DB-9 Connector. If the only available serial port on your computer is a DB-25, the following code will work; however, the pin numbers should be changed to reflect the pin-out of a DB-25 DTE connector. It should also be noted that the following project is available in completed form on the CD-ROM accompanying this book.

The basic controls are sufficient for this project—with one exception: a serial port must be accessed. Microsoft has an ActiveX control available called **MSComm** that can nicely handle this task. You can easily add this control to your Toolbar. Click "Project" on the Menu bar. Select "Components" from the pull-down menu. In the "Component Window," use the vertical scroll bar to search the list box for "Microsoft Comm control 6.0." The components are arranged alphabetically. Place a check on the checkbox for "Microsoft Comm control 6.0." Another way to add a component is by right-clicking on the Toolbox in the pop-up dialogue box and then selecting "components." Follow the same steps to remove an unwanted component by unchecking the associated checkbox.

Once you exit Components by clicking on the OK button, a new icon of a telephone sitting on top of a modem should appear in your control Toolbox. If you place your mouse cursor over it, "MSComm" should appear as ToolTipText. Press on the MSComm icon and add this control to the form by pressing down with the left mouse button. Then, with a diagonal sweeping motion, create a box. Once you release the left button, the MSComm icon will appear on the form. The default properties are acceptable for this project. You should take the time to examine the various properties of MSComm. Highlight the property by left-clicking on it and press function key F1 to retrieve Help information. The position of the MSComm control on the form is not critical, and it won't be visible at run time.

Next, locate the Frame Control in the Toolbox and add a Frame Control to the form that is roughly 1 inch high by 3 inches long. Change the caption on Frame to read "Control Outputs." The Frame Control is used in this application for grouping and identification. It's not necessary to change any properties. The form should look like Figure 2.08.

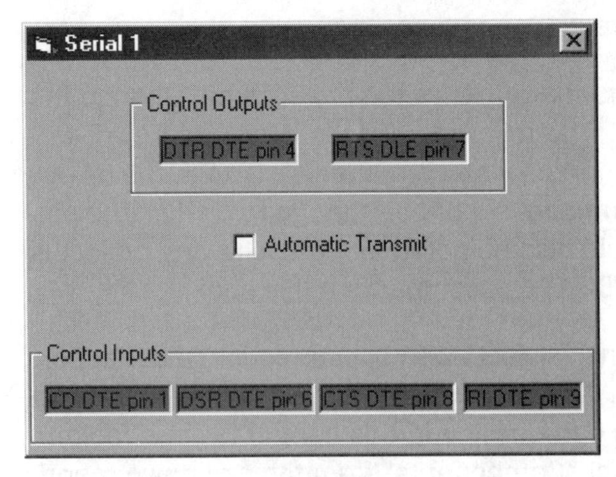

Figure 2.08

Click on a Label control and place it on the left-hand side of Frame1. Change the (name) property to "lblDTR". The lbl is following Microsoft's recommended naming convention. Change "lblDTR" caption to "DTR DTE pin 4". Change lblDTR AutoSize to true. Change lblDTR BorderStyle to "1 - Fixed Single". Change the ToolTipText to "Click on to toggle state." Add another label to Frame1 directly to the right of lblDTR and change the (name) property to "lblRTS".

The additional required property changes to lblRTS are as follows.

```
Caption= RTS DLE pin 7
AutoSize= true
BorderStyle= 1 - Fixed Single
ToolTipText= Click on to toggle state
```

These two labels complete the control output group. Click on the components and arrange them to provide symmetry in the Frame1 container. Adjust the size of Frame1 until everything looks good. Notice that these labels are now part of the Frame1 container. If you move Frame1, the labels move accordingly. Change the caption of frame 1 to read "Control Outputs."

At this point, you still have the control inputs and the transmit-out character to implement, but you can do a little coding to check out what you've accomplished so far. Double-click on Form1 to activate the code window. Place the following code into the Form Load event.

```
Private Sub Form_Load()
 MSComm1.CommPort = 2 ' Place your comport here 1=com1; 2=com2
 ' 9600 baud, no parity, 8 data, and 1 stop bit.
 MSComm1.Settings = "9600,N,8,1"
 ' Tell the control to read entire buffer when Input
 ' is used.
 MSComm1.InputLen = 1
 ' Open the port.
 MSComm1.PortOpen = True
 ' Initialize DTR and RTS to OFF
 MSComm1.DTREnable = False
 MSComm1.RTSEnable = False
 'use QBColor to set label colors
 lblDTR.BackColor = QBColor(12) '12=red 10=green
 lblRTS.BackColor = QBColor(12) '12=red 10=green
End Sub
```

In order to ensure that the Com port is turned off when the program is finished, place the MSComm1.PortOpen = false statement in the Form_Unload event.

```
Private Sub Form_Unload(Cancel As Integer)
MSComm1.PortOpen = False
End Sub
```

Next, add the output mouse-click functions. Both outputs are controlled by software that simply changes the state for each output as the label control is clicked on. The following sub Procedures lblRTS_Click and lblDTR_Click located below contain the required software. The "True" command is analogous to "on." According to the electrical characteristics, the control output should be a positive number greater than 3 volts DC when on and a negative number less than -3 volts DC when off. The label animation software will color the label red in the off state and green in the on state. The QBColor keyword is used for this task. You can change these colors to any combination you desire.

```
Private Sub lblRTS_Click()
 If MSComm1.RTSEnable = True Then
 MSComm1.RTSEnable = False: lblRTS.BackColor = QBColor(12) 'color red
 Else: MSComm1.RTSEnable = True: lblRTS.BackColor = QBColor(10) 'color green
 End If
End Sub


Private Sub lblDTR_Click()

 If MSComm1.DTREnable = True Then
  MSComm1.DTREnable = False: lblDTR.BackColor = QBColor(12) 'color red
 Else: MSComm1.DTREnable = True: lblDTR.BackColor = QBColor(10) 'color green
 End If

End Sub
```

Hook up your serial connector test jig and set up your digital multimeter or oscilloscope. Place the negative of your scope or meter on the common (pin 5) and run the program. Blow away the smoke and take a reading—just joking! As you click on each output label, you should see the label change color and the associated pin change state. If you don't, then verify that you're working with the correct Com port number and double-check your test setup for crossed wires, opened wires, bad solder connections, and so on. Well, that was easy—a few minutes of programming and you already have two programmable outputs to the outside world.

Now for the programming task required for the serial port inputs. As with the control outputs, use a Frame Control to hold the labels that you will be using to represent the status of the serial port control inputs: CD, DSR, CTS, and RI. Place the new Frame Control somewhere toward the bottom of the Form, size accordingly, and change the frame caption to read "Control Inputs."

Add a label control to the left side of Frame2 and change the (name) property to "lblCD". Additional required property changes to lblCD are as follows.

```
Caption= CD DTE pin 1
AutoSize= true
BorderStyle= 1 - Fixed Single
```

Add another label control to Frame2 to the right of lblCD and change the (name) property to "lblDSR". Additional required property changes to lblDSR are as follows.

Caption= DSR DTE pin 6
AutoSize= true
BorderStyle= 1 - Fixed Single

Add another label control to Frame2 to the right of lblDSR and change the (name) property to "lblCTS". Additional required property changes to lblCTS are as follows.

Caption= CTS DTE pin 8
AutoSize= true
BorderStyle= 1 - Fixed Single

Add the last label control to Frame2 to the right of lblCTS and change the (name) property to "lblRI". Additional required property changes to lblRI are as follows.

Caption= RI DTE pin 9
AutoSize= true
BorderStyle= 1 - Fixed Single

Programming of the inputs requires the MSComm OnComm event. On the code section of the form, open up the control list box and scroll down to MSComm. The Private Sub MSComm1_OnComm() should now be added to your code. Highlight the term OnComm with the mouse and press function key F1 to activate Help. The OnComm event is triggered whenever a communication event or an error occurs. Click on the example link in the Help window. A properly structured Select Case is there for the taking. Copy the Select Case code and paste it into the program code for MSComm_OnComm. Modify this code as illustrated in **bold** below with the input-control animation.

```
Select Case MSComm1.CommEvent
' Handle each event or error by placing code below each case statement

' Errors
Case comEventBreak ' A Break was received
Case comEventFrame ' Framing Error
Case comEventOverrun ' Data Lost
```

```
Case comEventRxOver ' Receive buffer overflow
Case comEventRxParity ' Parity Error
Case comEventTxFull ' Transmit buffer full
Case comEventDCB ' Unexpected error retrieving DCB]
' Events
Case comEvCD ' Change in the CD line.
  If MSComm1.CDHolding = False Then
  lblCD.BackColor = QBColor(12) 'red
 Else: lblCD.BackColor = QBColor(10) 'green
 End If

Case comEvCTS ' Change in the CTS line.
   If MSComm1.CTSHolding = False Then
  lblCTS.BackColor = QBColor(12) 'red
 Else: lblCTS.BackColor = QBColor(10) 'green
 End If

Case comEvDSR ' Change in the DSR line.
  If MSComm1.DSRHolding = False Then
   lblDSR.BackColor = QBColor(12) 'red
  Else: lblDSR.BackColor = QBColor(10) 'green
  End If

Case comEvRing ' Change in the Ring Indicator.
  If lblRI.BackColor = QBColor(10) Then
  lblRI.BackColor = QBColor(12)  'red
  Else: lblRI.BackColor = QBColor(10) 'green
  End If

Case comEvReceive ' Received Rthreshold # chars.
Case comEvSend ' There are SThreshold number of characters in the
               ' transmit buffer
Case comEvEOF ' An EOF character was found in the input stream
End Select
```

Before you run this code, the control inputs need to be initially colorized, so add the following code to the bottom of the Form_Load event.

```
' control input initialization
lblCD.BackColor = QBColor(12) 'red
lblCTS.BackColor = QBColor(12) 'red
lblDSR.BackColor = QBColor(12) 'red
lblRI.BackColor = QBColor(12) 'red
```

When you run the program, all the status lights for the control signals are colored red. The initialization code found in the Form_Load event accomplishes this task. At this point, all of the control inputs are floating, since they are not connected to an output. The interface integrated circuit that performs the RS-232 voltage translation for the computer contains roughly a 5000-ohm resistor on each input tied to common. This pull-down resistor ensures that the control inputs—and the Receive Data input, for that matter—are not arbitrarily bouncing around or floating when nothing is connected. In order for a control input to be considered on, the input must exceed +3 volts. If you turn on either DTR or RTS, a positive voltage is now available on that pin. Use a jumper to route this positive voltage to each control input. Each status indicator should change from red to green to red as you go from input to input.

The final task for this project is to create a periodic asynchronous byte output on the TD line. This periodic output will make it easy for an oscilloscope to sync up for a visual inspection of the RS-232 byte transmission. If you do not have a scope, you will see a change in potential with your multimeter when the byte is being transmitted. A checkbox control will be used to enable/disable this function.

Add a timer control to the form. Set the timer-interval property to 10. This will cause the timer-event property to be executed approximately every 10 milliseconds. At 9600 Baud, the width of each bit is equal to 104.16 microseconds (1/9600). The MSComm setting property for this project is 9600,N,8,1. Therefore, the transmission time for a byte is 1.0416 milliseconds (10 * 1/9600). This timing is well within the 10-millisecond interval specified for the timer.

Add a checkbox control to the form and position it to your liking. Change the caption property for Check1 to "Periodic Transmission". Adjust the sizing of the Check1 checkbox control with your mouse so that the caption is arranged on one line. Now add the following code to the Timer1_Timer event.

```
Private Sub Timer1_Timer()
'An inspection of Check1 property value specifies that
'0= unchecked and 1 = check
  'if Check1 value is equal to 1 indicating that the box is checked then transmit
  'otherwise do nothing
If Check1.Value = 1 Then Let MSComm1.Output = Chr(1)

End Sub
```

The Chr keyword returns a string containing the character associated with the specified ASCII character code. This keyword instructs MSComm1 to transmit a byte containing 01 hex and not the ASCII character 1, which is equivalent to 31 hex.

Run the program, and with the mouse place a check on the Check1 checkbox. If you have an oscilloscope, then adjust the time base to 100 microseconds per grid division. Since oscilloscopes are configured with 10 horizontal grids, you can almost see all 10 bits of the transmitted data byte on the oscilloscope's screen. Approximately 40 microseconds are spilled off the end of the scope's display, because each bit time is 104.16 microseconds.

Figure 2.09 illustrates an oscilloscope view of the RS-232 voltage-translated transmitted data byte. A voltage of less than –3 volts represents a mark (logic 1) on the data line. A voltage of +3 volts or more indicates a space (logic 0) on a data line. The logic level for each bit is included in the illustration.

If you are using an oscilloscope, you should experiment by outputting different characters, Bauds, parities, and Stop bits. If you drop down into the slower Baud rates, make sure that you adjust the timer-interval rate to an appropriate value. At 110 Baud, the data character requires a transmission time of 90.9 milliseconds. If the timer rate were to remain at the 10 milliseconds interval value, the transmitted bytes begin running together and make it difficult for the oscilloscope to sync up. As an exercise, you may want to add a receive section to this program. The transmitted data can be looped back into the receive input, RD. Use a Textbox control to display the received data as an ASCII character, a decimal value, or a hexadecimal value. The MSComm input property is used to pull characters from the receive buffer.

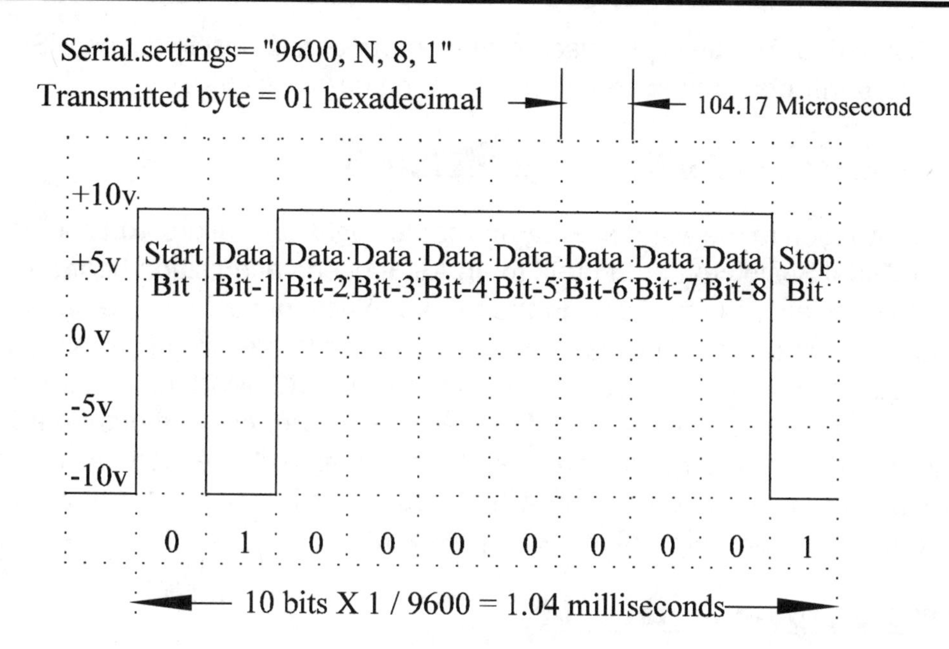

Figure 2.09

No doubt you will agree that this first programming task was accomplished with ease. After working with the signals available on the RS-232 connector, the functions and characteristics of these signals should be clearer. The data-flow control signals are really only necessary when dealing with the transfer of large blocks of data. Actually, software data-flow control could also be used, thereby eliminating the need for hard-wired data-flow control. The serial port's control inputs and control outputs with proper electrical conditioning could be used to interface with external devices such as switches and relays. Each serial port has two outputs and four inputs that could be used to interface with external devices.

The MSComm control is a very powerful tool to have in the Toolbox. MSComm's event-driven communications is a very efficient technique for handling serial port interactions. Although you now have explored the primary properties of the MSComm control, you should also be aware of other properties that are available. You should step through each property of MSComm. Use the F1 function key to retrieve additional information on each property. The transmit buffer (OutBufferSize) and receive buffer (InBufferSize) can be adjusted. Four modes of handshaking can be specified. The ability to discard null characters is possible. The MSComm control certainly does make serial programming easy.

Second Serial Communication Program

Before moving on to the world of Programmable Logic Controllers, you should construct a Visual Basic program to talk to an ASCII-based acquisition board. A number of these boards are available in the market place, or you can use one of your own homegrown designs. Typically, acquisition boards are microcontroller-based circuit boards. The microcontroller uses an Intel 80C32, Motorola, Microchip, or some other Integrated circuit manufacturer's CPU core. Acquisitions boards are used to read temperature, pressure, and other real-world sensors. A host computer polls a request, and the acquisition board sends a serial ASCII string with the information. The serial link is usually RS-232.

General Purpose Serial Link

The software objective of this program is:

1) Provide a user interface where communication parameters are selectable and save these parameters to a configuration file.
2) Provide a user interface where data-acquisition parameters are selectable and save these parameters to a configuration file.
3) Data log to file.

KT32 is the data-acquisition circuit board being used as an example in this project. This board, among other functions, provides temperature information in degrees Fahrenheit. The interface is RS-232, with a data-byte structure of 9600N81.

KT32 Read Temperature Command
$T<carriage return>
RESPONSE: @+0074.63< carriage return >
NOTE: + IS SIGN OF TEMPERATURE IN DEGREES F
@-0159.69< carriage return > INDICATES AN OPEN PROBE
@+0140.18< carriage return > INDICATES A SHORTED PROBE

This software will be developed to be universal, as far as interfacing to serial devices of this type where the protocol requires a simple request/response format without error checking. The command field will be user programmable, with the command stored in a configuration file. Therefore, upon activation of the program, the previously set parameters will be loaded. In addition, a response field will be included and programmable to ensure that the received data fits the profile of the device.

Now, start a new project and call it "GPSL," which in this case stands for "General Purpose Serial Link." Launch Visual Basic and select Standard EXE. Save the project as "GPSL." Make sure that you select the appropriate folder for your project.

The first task is to set up a user interface for configuring the desired communication parameters. Since this task is secondary to the data-acquisition part of the project, another form will be added to the project. By selecting "Add Form" in the Project menu, and then selecting "Form" in the subsequent dialogue box, you create the additional Form. You should see a second form in your project browser called "Form2(form2.frm)". Change the name of the form to "Comparams". Place "Communication Parameters" into the forms-caption property.

The communication parameters are all standards and predefined. The communication ports are 1 and 2. The Baud rates are standard rates. The number of stop bits, parity, and the number of data bits are all predefined. The best tool for this task is the Option button. Option buttons are automatically linked to each other within a container so that when one is selected, all the other Option buttons within the container are deselected. This feature minimizes the required software for this configuration task.

There are five groups of parameters that need to be configured: com port, Baud rate, parity type, the number of Data bits, and the number of Stop bits.

Since the Option buttons are container interrelated, additional containers are required to provide isolation for the selection task. Otherwise, if you placed all of the required Option buttons on just the form container, only one Option button could be selected. The Frame Control is ideal for the isolation task. In addition, when the frame is moved, the associated Option buttons follow suit.

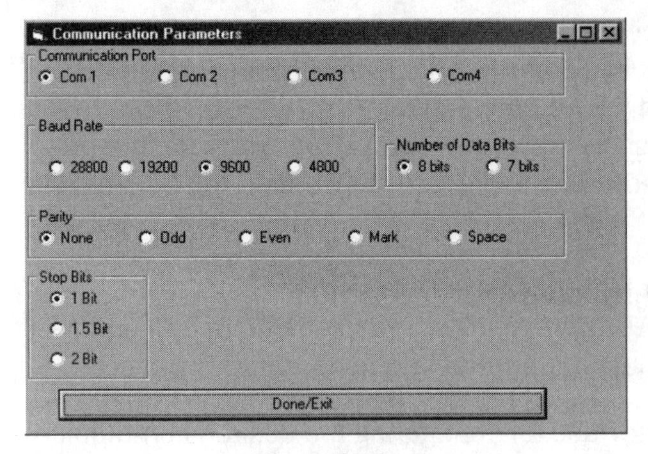

Figure 2.10

Place a Frame Control on the "Comparams" form. Visual Basic will automatically set the name of the frame to Frame1. Since the frame task is simply a container, this name won't be altered. Change the Frame1 caption to read "Communication Port". Place an Option button into Frame1 and name it "optCOM1". Add three more Option buttons into Frame1 and name them "optCOM2", "optCOM3", and "optCOM4". Change the caption for each of these Option buttons accordingly. The CommPort Property can be any number from 1 to 16.

Add another Frame Control ("Frame2") to the "Comparams" form and change the caption to "Baud Rate". Place four Option buttons into Frame2 and name them "optBR288K", "optBR192K", "optBR96K", and "optBR48K". Change the respective captions to 28800, 19200, 9600, and 4800.

Add a third Frame Control ("Frame3") to the "Comparams" form and change the caption to "Parity". Place five Option buttons into Frame3 and name them "optPNONE", "opt optPODD", "optPEVEN", "optPMARK", and "optPSPACE". Change the respective captions to None, Odd, Even, Mark, and Space.

Add yet another Frame Control ("Frame4") to the "Comparams" form and change the caption to "Number of Data Bits". Place two Option buttons into Frame4 and name them "optDATA8" and "optDATA7". Change the respective captions to 8 bits and 7 bits. MSComm support bit sizes of 5, 6, 7, and 8. Seven-bit and 8-bit data sizes are the most common.

Add one more Frame Control ("Frame5") to the "Comparams" form and change the caption to "Stop Bits". Place three Option buttons into Frame5 and name them "optBIT1", "optBIT15", and "optBIT2". Change the respective captions to 1 Bit, 1.5 Bit, and 2 Bit.

Finally, add a Command button named "cmdDONEexit" with a caption of "Done/Exit". You can arrange these various groups of Option buttons to your liking. The book arrangement is depicted in Figure 2.10.

Temporarily add the following code shown in **bold** to the Form load procedure of Form1. The two statements will immediately launch Form2 when the project is started. Start the project, click the various Option buttons from group to group, and verify that only one Option button is active per group. An active Option button's value property is True. The value property of all the inactive Option buttons for each group is false. During run time, the value property for each Option control can be either interpreted or set. Remove the temporary code when you are finished.

```
Private Sub Form_Load()
Comparams.Enabled = True
Comparams.Visible = True

End Sub
```

As you can see from the previous exercise, the Option-button control makes the programming task of setting the communication parameters very easy. Once these parameters are set and the communication link is successful, there is no reason for this form to be displayed. The parameters will be saved to a configuration file, which will automatically get loaded on program start-up. A control will be placed on Form1 to activate Form2, in order to facilitate the changing of the communication parameters. When the form is closed, the MSComm1 communication parameters and the configuration file will be updated.

Since the communication variables need to be available for both forms, a module should be added to the project. On the Menu bar, select Project/Add Module/New/Open. Notice that the new module is added to the Project window. Leave the module name as "Module1". Within this module, the communication variables can be declared as Public and bridge the gap between forms. In addition, modules are good places to store code procedures that may be utilized in other projects.

In the general-declaration section of the recently added module, place the following statements.

```
Option Explicit

'This variable sets the Communication port
Public sComPort As String
'This variable sets the Baud rate.
Public sBaudRate As String
'This variable sets the parity type.
Public sParity As String
'This variable sets the number of data bits.
Public sDataBits As String
'This variable sets the number of stop bits.
Public sStopBits As String
```

In the Module1, add Sub procedures ReadComParamFile and WriteComParamFile. ReadComParamFile opens a file named "ComPar.txt" and retrieves the communications parameters into Visual Basic variables for insertion into MSComm. An error handling routine is provided for a "File Not Found" error in order to load default communications parameters and create the "ComPar.txt" file. The "txt" extension makes the file readily accessible for inspection with Microsoft's Notepad text editor. If any other error is encountered during the file read, a message box with the error type is launched and the program is subsequently ended. WriteComParamFile writes the communication parameters to a file called "ComPar.txt".

The program flow of these procedures can be observed by placing the ReadComParamFile statement in the Form_Load procedure of Form1. Place a Breakpoint on this statement. Run the program. The Breakpoint will immediately stop execution of the program. Now you should single step through the code using the F8 function key. Place the mouse pointer over the variables to examine the content.

```
Public Sub ReadComParamFile()
Dim FILENUM As Byte 'temporary scratch pad variable
Dim A As String 'temporary scratch pad variable

FILENUM = FreeFile
```

```
'Look for a file called "ComPar.txt"
On Error GoTo HandleError

Open "ComPar.txt" For Input As #FILENUM
Input #FILENUM, A
'use Trim command to trim out leading and trailing spaces.
sComPort = Trim(A) 'TRIM OUT ANY SPACES
 Input #FILENUM, A
 sBaudRate = Trim(A) 'TRIM OUT ANY SPACES
 Input #FILENUM, A
 sParity = Trim(A) 'TRIM OUT ANY SPACES
  Input #FILENUM, A
  sDataBits = Trim(A) 'TRIM OUT ANY SPACES
  Input #FILENUM, A
  sStopBits = Trim(A) 'TRIM OUT ANY SPACES
 Close FILENUM
 Exit Sub 'Exit procedure; task complete
HandleError: 'error handler
 Select Case Err.Number ' Evaluate error number error object
  Case 53 ' "File Not Found"
   Close #FILENUM ' Close open file
   'Load default communication parameters
   sComPort = "1"
   sBaudRate = "9600"
   sParity = "N"
   sDataBits = "8"
   sStopBits = "1"
   WriteComParamFile 'write parameters to file
  Case Else
   'this path for all errors other than File not Found
   'Display the type of error and end program!!!
   Close #FILENUM
   MsgBox Err.Description 'provides a description
   End 'End program
   End Select
 End Sub
```

```
Public Sub WriteComParamFile()
Dim FILENUM As Byte
Dim A As String
 FILENUM = FreeFile
 A = sComPort + "," + sBaudRate + "," + sParity + ","
 A = A + sDataBits + "," + sStopBits
 Open "ComPar.txt" For Output As #FILENUM
 Print #FILENUM, A
 Close #FILENUM
End Sub
```

Run Microsoft's Notepad to view the contents of the file. Figure 2.11 shows the contents of ComPar.txt when viewed with Notepad. Now write the code for Form2. This form makes the communication parameters user-selectable. Place the following statements in the Form_Load procedure of Form2. In the form-load event, the value for each option-button group is set according to the communication parameters. The communication parameters were previously extracted from the "ComPar.txt" file.

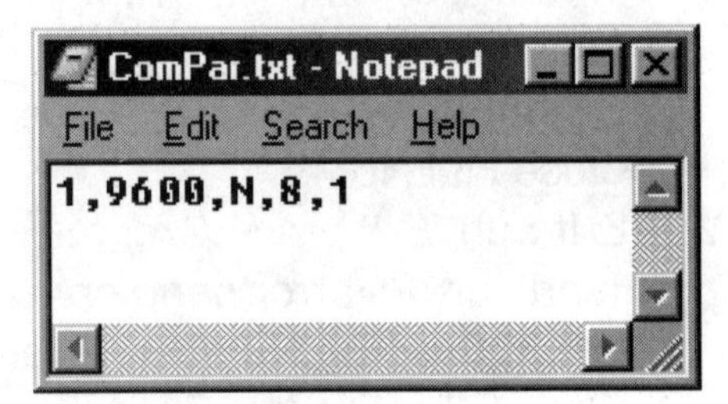

Figure 2.11

```
Private Sub Form_Load()

If sComPort = "1" Then optCom1.Value = True
If sComPort = "2" Then optCom2.Value = True
If sComPort = "3" Then optCom3.Value = True
If sComPort = "4" Then optCom4.Value = True
 If sBaudRate = "28800" Then optBR288K.Value = True
 If sBaudRate = "19200" Then optBR192K.Value = True
 If sBaudRate = "9600" Then optBR96K.Value = True
 If sBaudRate = "4800" Then optBR48K.Value = True
 If sParity = "N" Then optPNONE.Value = True
 If sParity = "O" Then optPODD.Value = True
 If sParity = "E" Then optPEVEN.Value = True
 If sParity = "M" Then optPMARK.Value = True
```

```
If sParity = “S” Then optPSPACE.Value = True
If sDataBits = “8” Then optDATA8.Value = True
If sDataBits = “7” Then optDATA7.Value = True
If sStopBits = “1” Then optBIT1.Value = True
If sStopBits = “1.5” Then optBIT15.Value = True
If sStopBits = “2” Then optBIT2.Value = True
End Sub
```

The Form_Unload event of Form2 sets the communication parameters according to the Option buttons selected for each communications group. These values are also saved to the “ComPar.txt” file. Place the following code in Form2’s Form_Unload event.

```
Private Sub Form_Unload(Cancel As Integer)
If optCom1.Value = True Then sComPort = “1”
If optCom2.Value = True Then sComPort = “2”
If optCom3.Value = True Then sComPort = “3”
If optCom4.Value = True Then sComPort = “4”
If optBR288K.Value = True Then sBaudRate = “28800”
If optBR192K.Value = True Then sBaudRate = “19200”
If optBR96K.Value = True Then sBaudRate = “9600”
If optBR48K.Value = True Then sBaudRate = “4800”
If optPNONE.Value = True Then sParity = “N”

If optPODD.Value = True Then sParity = “O”
If optPEVEN.Value = True Then sParity = “E”
If optPMARK.Value = True Then sParity = “M”
If optPSPACE.Value = True Then sParity = “S”
If optDATA8.Value = True Then sDataBits = “8”
If optDATA7.Value = True Then sDataBits = “7”
If optBIT1.Value = True Then sStopBits = “1”
If optBIT15.Value = True Then sStopBits = “1.5”
If optBIT2.Value = True Then sStopBits = “2”

‘save new parameters to the communication configuration file
WriteComParamFile
End Sub
```

On Form1, add an MSComm control. The MSComm is not in the standard toolbox. Right-click on the toolbox and select components. Look for Microsoft Comm Control 6.0 and then click on the associated checkbox. MSComm is now located in the Toolbox. Place MSComm on the form. MSComm is not visible at run time, so form placement is not critical.

Add a Command button to Form1 named "cmdChangeComParam". Change the caption property of the command button to read, "Press to Change Communication Parameters." Place the command button at the bottom of Form1 and change the size of the button so the entire caption field is visible. Change the caption of Form1 to read, "General Purpose Serial Link". The form should look like Figure 2.12.

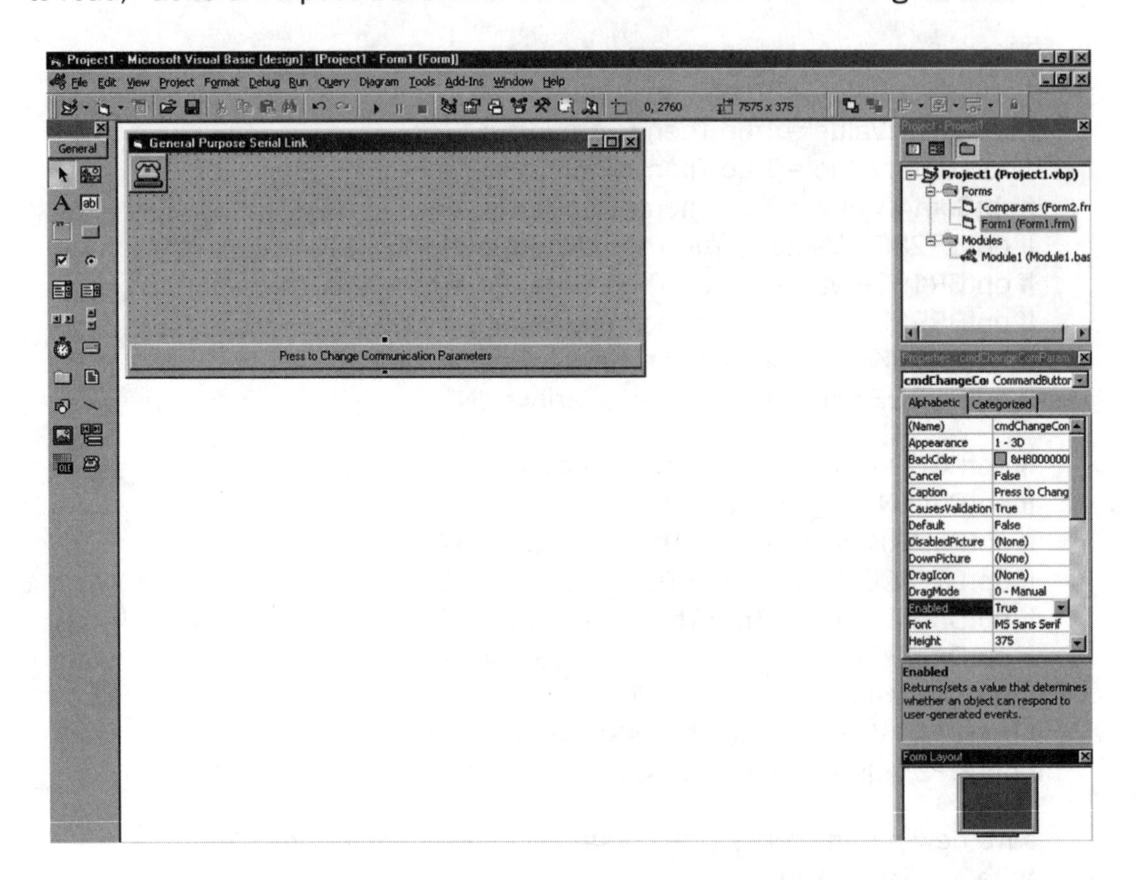

Figure 2.12

Add the following code to the cmdChangeComParam_Click() event.

```
Private Sub cmdChangeComParam_Click()
Form1.Visible = False
Comparams.Visible = True
Comparams.SetFocus
End Sub
```

This code makes Form1 not visible and changes the visibility and focus of the Comparams Form. Make sure that you add the statement Form1.Visible =True to the Comparams Form_Unload event. Otherwise, when you close out the communication parameters window or Form2, none of the project windows will be visible. You'll be looking at the desktop. Run the program and navigate from one window to the next. Change the options as you go and verify that the options are maintained.

The next task requires data collection and storage. Since this program is targeted to be of a general-purpose nature, the ability to enter the required command must be provided. Additionally, a data-configuration file is required to store this command so it won't be necessary to reenter the command the next time the program is run.

Data-log capability is also required where, at a specified time interval, the command is automatically issued. The response from the connected serial device is collected and stored away to a file along with a time stamp and a data stamp. Commas will be used to delimit the data and time/date information. The data-log file name will be "Data[mmm][yyyy].csv". The bracket [mmm] indicates the first three alpha-characters of the current month. The bracket [yyyy] represents a numerical description of the current year. Csv represents a standard file extension designation, indicating that the file consists of comma-separated variables. All user-specified data-log parameters will be stored to the data-configuration file.

Add a Command button to Form1 and change the name to cmdSend. The caption for cmdSend is "Send Command".

Add a label called "lblCommand" to Form1 next to the "Send Command" command button. Set the AutoSize property to True and the BorderStyle to "1- Fixed Single". In the ToolTipText property, add the following text: "Double Click to enter a new Command".

Directly below the "lblCommand" label, place a Textbox control. The Textbox should be named "tbCommand".

Place another label control below the textbox. The name of this label is "lblReceivedData". Set the AutoSize property to True and the BorderStyle to "1-Fixed Single". Place a Checkbox on Form1 and name it "cbDataLog". Change the caption to read "Auto DataLog".

Now you need to add several Option buttons to select the data-log time interval. First, you will need to add a frame to act as a container for the Options buttons. This frame simply serves as a container, so you can stay with the default name. Set the caption property of Frame1 to "Data Log Time Interval". Now add four Option buttons to Frame1. The names of these Option buttons are opt5Min, opt10Min, opt15Min, and opt30Min. Adjust the caption for each of these Option buttons accordingly. A layout of Form1 is illustrated in Figure 2.13.

Three setting are required to be stored in the data-configuration file. They are the command, the status of Auto Data-log, and the data-log time interval. The name of the configuration file is "DataPar.txt". Actually, the code is almost identical to the code written to set/retrieve the communication parameter. The exceptions are the name of the file, the variable names, and the number of variables. If you were keying in this code, the best method would be to copy, paste, and modify both ReadComParamFile and WriteComParamFile.

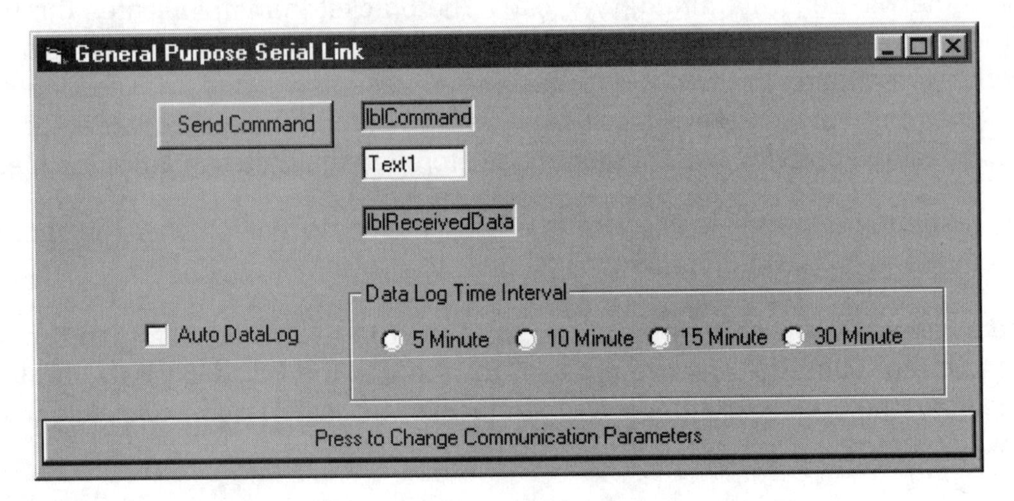

Figure 2.13

Place the following variable declaration to the general-declaration procedure of Module1.

```
'Variable for Data Log Settings
'This variable stores the command
Public sCommand As String
'This variable stores the whether data log is enabled
Public sDataLogStatus As String
'This variable store the Data Log time interval
Public sDataLogTime As String
```

Add the following code to Module1.

```
Public Sub ReadDataParamFile()
Dim FILENUM As Byte 'temporary scratch pad variable
Dim A As String 'temporary scratch pad variable

FILENUM = FreeFile
'Look for a file called "DataPar.txt"
On Error GoTo HandleError

Open "DataPar.txt" For Input As #FILENUM
Input #FILENUM, A
'use Trim command to trim out leading and trailing spaces.
sCommand = Trim(A) 'TRIM OUT ANY SPACES
 Input #FILENUM, A
 sDataLogStatus = Trim(A) 'TRIM OUT ANY SPACES
 Input #FILENUM, A
 sDataLogTime = Trim(A) 'TRIM OUT ANY SPACES

Close FILENUM
Exit Sub
HandleError: 'error handler
 Select Case Err.Number ' Evaluate error number using error object
  Case 53 ' "File Not Found"
   Close #FILENUM ' Close open file.
   'Load default Data Log parameters
```

```
   sCommand = "$T"
   sDataLogStatus = "True"
   sDataLogTime = "5"
   WriteDataParamFile 'write parameters to file
  Case Else
   'this path for all errors other than File not Found
   'Display the type of error and end program!!!
   Close #FILENUM
   MsgBox Err.Description 'provides a description
   End 'End program
 End Select
End Sub
```

Add another procedure to Module1 called "WriteDataParamFile".

```
Public Sub WriteDataParamFile()
Dim FILENUM As Byte
Dim A As String
 FILENUM = FreeFile
 A = sCommand + "," + sDataLogStatus + "," + sDataLogTime
 Open "DataPar.txt" For Output As #FILENUM
 Print #FILENUM, A
 Close #FILENUM
End Sub
```

Add the following code to the Form1 Form_Load procedure after the ReadComParamFile statement.

```
'Get Data Log Parameters
ReadDataParamFile
'Initialize Project check box and option buttons
 If sDataLogStatus = "True" Then cbDataLog.Value = 1 'place check if true
 If sDataLogTime = "5" Then opt5Min = True
  If sDataLogTime = "10" Then opt10Min = True
  If sDataLogTime = "15" Then opt15Min = True
   If sDataLogTime = "30" Then opt30Min = True
 ' Show the command value
```

```
lblCommand.Caption = "Command => " & sCommand
'hide text box
tbCommand.Visible = False
'null receive label
lblReceivedData.Caption = "Received Data =>"
```

Once the following procedures are added to the Form1 code, the user interface is complete. You can run the project and change any parameter—whether it is communications options or data-log options, the selection is saved to the respective files.

```
Private Sub cbDataLog_Click() 'CheckBox
If cbDataLog.Value = 0 Then
 sDataLogStatus = "False"
Else: sDataLogStatus = "True"
End If
 WriteDataParamFile
End Sub
```

```
Private Sub opt5Min_Click() '5Min Option button
sDataLogTime = "5"
 WriteDataParamFile
End Sub
```

```
Private Sub opt10Min_Click() '10Min Option button
sDataLogTime = "10"
 WriteDataParamFile
End Sub
```

```
Private Sub opt15Min_Click() '15Min Option button
sDataLogTime = "15"
 WriteDataParamFile
End Sub
```

```
Private Sub opt30Min_Click() '30Min Option button
sDataLogTime = "30"
 WriteDataParamFile
End Sub
```

```
Private Sub lblCommand_DblClick() 'Command label
 tbCommand.Visible = True
 tbCommand.SetFocus
  tbCommand.Text = ""
End Sub
```

```
'TextBox key Press look for a carriage return
Private Sub tbCommand_KeyPress(KeyAscii As Integer)
'Command entry completion is a carriage return character
 If KeyAscii = Asc(vbCr) Then
 'if only a carriage return was keyed into the textbox then close out
 'and leave the existing command intact
 If tbCommand.Text <> "" Then sCommand = tbCommand.Text
  WriteDataParamFile 'save to file
  tbCommand.Visible = False 'hide textbox
   'display new command
   lblCommand.Caption = "Command => " & sCommand
 End If
End Sub
```

The following code sets up the communications and the data-logging aspects of the project. Essentially, there are two elements issuing information requests to the serial device: the manual command button, and the data logger. Add the following to the general declaration procedure in Module1.

```
'Variable for Data Log Settings
'This variable stores the command
Public sCommand As String
'This variable stores the data log status (enabled/disabled)
Public sDataLogStatus As String
'This variable store the Data Log time interval
Public sDataLogTime As String

'Used for DataLog
Public LogDataFlag As Boolean
Public DataLogFlag As Boolean
'Used for communication
```

Public CommandSentFlag As Boolean

Add the following as additional procedures to Form1.

```
Private Sub cmdSend_Click()
cmdSend.Tag = 1 'send command request
' take notice that cmdSend.Tag is a property of the Command button
that is utilized for this task
'tmrCommunication_Timer will route the request to MSComm
End Sub
```

```
Private Sub tmrCommunication_Timer()
 Dim vRTseconds As Variant
 Dim sRTmin As String
 'The Now keyword provides real time information
 vRTseconds = Second(Now)
 sRTmin = Minute(Now)

 'Check automatic datalog issued commands before the manual request
If DataLogFlag = True And vRTseconds > 5 Then DataLogFlag = False

If sDataLogStatus = "True" And CommandSentFlag = False Then
 'check datalog time Every 5 minute interval
  If DataLogFlag = False And vRTseconds < 5 Then
 If sDataLogTime = "5" And (Right(sRTmin, 1) = "0" Or Right(sRTmin,
1)="5") Then
   LogDataflag = True: DataLogFlag = True: SendCommand
  End If
   'every 10 minute interval
  If sDataLogTime = "10" And Right(sRTmin, 1) = "0" Then
    LogDataFlag = True: DataLogFlag = True: SendCommand
  End If
   'every 15 minute interval
  If sDataLogTime = "15" And _
  (sRTmin = "0" Or sRTmin = "15" Or sRTmin = "30" Or sRTmin = "45") Then
    LogDataFlag = True: DataLogFlag = True: SendCommand
  End If
```

```
  'every 30 minute interval
  If sDataLogTime = "30" And (sRTmin = "0" Or sRTmin = "30") Then
   LogDataFlag = True: DataLogFlag = True: SendCommand
  End If
  End If
End If
 'Manual send command path by "Send Command" button control
If CommandSentFlag = False And cmdSend.Tag = 1 Then
 SendCommand
 cmdSend.Tag = 0
End If
End Sub
```

```
Private Sub SendCommand()
 'Send Command through MSComm1.output
 CommandSentFlag = True
 MSComm1.Output = sCommand + vbCr
 tmrTimeOut.Interval = 60  '60 milliseconds
 tmrTimeOut.Enabled = True  'Start Timer
End Sub
```

```
Private Sub MSComm1_OnComm()
 'Collect and Route Response from the device. Datalog if specified
Static sRxData As String
Dim sRD As String
Select Case MSComm1.CommEvent

 Case comEvReceive  ' Received RThreshold # of chars.
 sRD = MSComm1.Input
 sRxData = sRxData & sRD
  If Asc(sRD) = Asc(vbCr) Then
   CommandSentFlag = False
   If LogDataFlag = True Then  'DataLog Path
    StripOffLeadingNonNumerics sRxData
     DataLog (sRxData)  'Save Data to a file
     LogDataFlag = False
```

```
    tmrTimeOut.Enabled = False 'Stop Timer
   End If
   sRxData = "" 'Clear sRxData for next data collection
   'Show data collected on Form1
  Else: lblReceivedData.Caption = "Received Data =>" + sRxData
  End If
 End Select
End Sub
```

```
Public Sub StripOffLeadingNonNumerics(ByRef sRxData)
'ByRef option sends back the result to original variable
Dim bIndex As Byte 'Used for indexing
Dim NumberFlag As Boolean 'Flag used to exit Do—Loop
bIndex = 1 'initial starting point
Do
Select Case Mid(sRxData, bIndex, 1) 'Mid String Function
 Case "0" To "9", "+", "-", "." 'Elements of a real number
  NumberFlag = True 'Flag used to exit Do—Loop
  sRxData = Mid(sRxData, bIndex, (Len(sRxData) - bIndex))
 Case Else
  bIndex = bIndex + 1
 End Select
Loop Until bIndex = Len(sRxData) Or NumberFlag = True
End Sub
```

```
Public Sub DataLog(data)
 Dim FILENUM As Byte 'temporary scratch pad variable
 Dim sMOyr As String
 Let sMOyr = Format(Now, "MMMYYYY") 'JAN2000

 'Set file attributes to Read Only when exiting therefore
 'make the attribute normal for appending data. Setting
 'the attribute will stop programs like Excel from taking
 'complete control of the file when it is opened for inspection
 'or graphing.
 On Error Resume Next
```

```
 SetAttr "Data" + sMOyr + ".csv", vbNormal
 FILENUM = FreeFile
 Open "Data" + sMOyr + ".csv" For Append As #FILENUM
 Print #FILENUM, Date$ + "," + Time$ + "," + data
 Close #FILENUM
 SetAttr "Data" + sMOyr + ".csv", vbReadOnly
End Sub
```

```
Private Sub tmrTimeOut_Timer()
 'Time out timer required otherwise the program could hang up
 'Clear all communication flags
 LogDataFlag = False
  CommandSentFlag = False
   DataLogFlag = False
    cmdSend.Tag = 0
  tmrTimeOut.Enabled = False 'Stop Timer
End Sub
```

The manual command button generates a request for display purposes only. A request command based on real time is automatically issued by the data-logger control and the data is saved to a disk file. The data-log request sends the request at the zero second of the interval time. In the event that the operating system may be busy with another task at that moment, provisions are made for a five-second data-collection window. Reliable data collection is well worth the additional logic. This may seem a little excessive, but if the operating system is initializing a Win modem or busy with scandisk, a one-second delay is possible on even the fastest machines.

You will notice that the issuance of the request command is controlled by the timer named "tmrCommunication." This timer is activated every 60 milliseconds, and all communication requests are routed through it. Each request is operated on a first-come, first-serve basis. If a data-log time interval match occurs, the appropriate flags are set and a request command is issued. If a manual request is already in progress, the steering logic ensures that the command will be issued upon completion of the current request.

A communication timer called "tmrTimeOut" is provided to ensure that the program does not sit there forever looking for a carriage return and ignoring all other requests.

This timer is enabled when the request command is issued. Subsequently, when a carriage return is received, the timer is disabled.

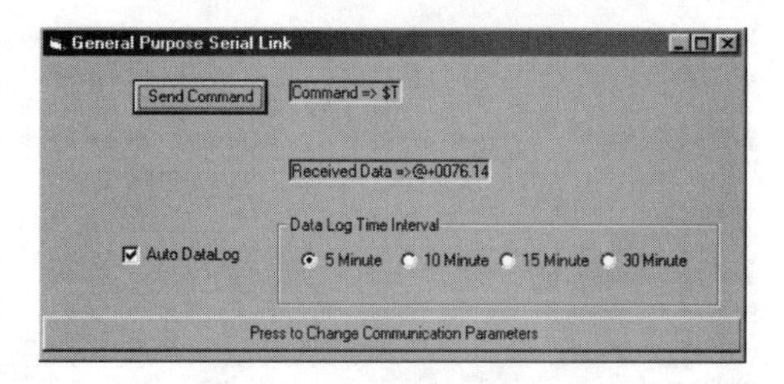

Figure 2.14

When constructing combinational logic statements such as in the Data-Log time-interval detection code, you should pay close attention to the grouping of the logic. For example, A and B or C or D will produce a completely different solution than A and (B or C or D). In the first case, the logic grouping is implicitly (A and B) or C or D. The second logic statement can be expanded as (A and B) or (A and C) or (A and D). As you can see, logical grouping is very critical to the desired result. Figure 2.14 illustrates an actual application with a KT32 serial temperature device. The Data-Log file with the csv extension is directly readable with Microsoft Excel. This is illustrated in Figure 2.15.

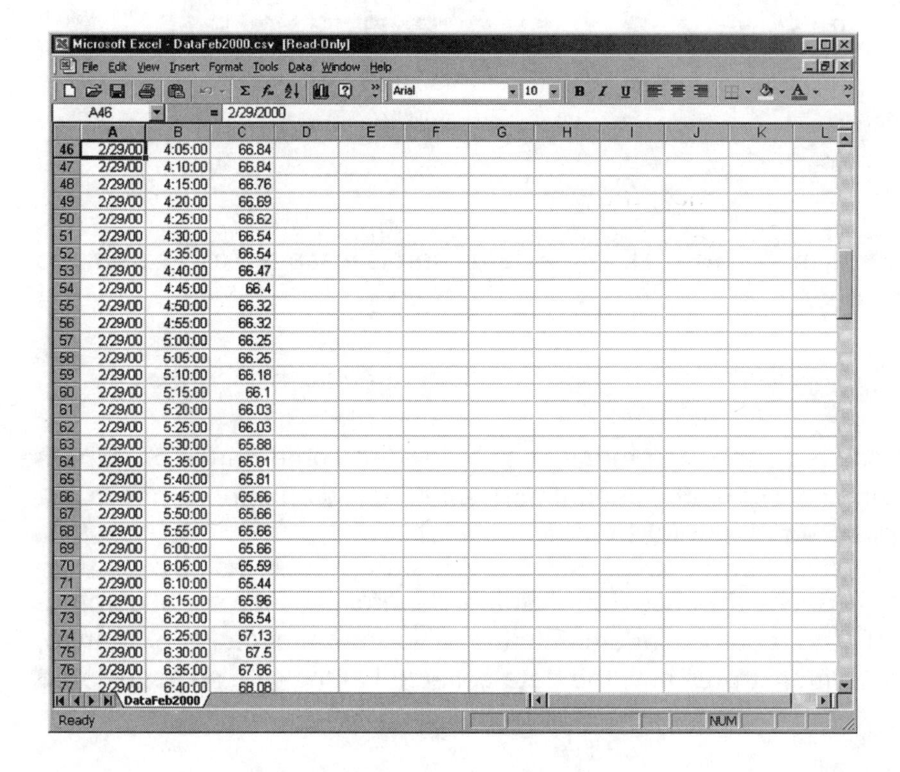
Microsoft Excel - DataFeb2000.csv [Read-Only]

A46 = 2/29/2000

	A	B	C
46	2/29/00	4:05:00	66.84
47	2/29/00	4:10:00	66.84
48	2/29/00	4:15:00	66.76
49	2/29/00	4:20:00	66.69
50	2/29/00	4:25:00	66.62
51	2/29/00	4:30:00	66.54
52	2/29/00	4:35:00	66.54
53	2/29/00	4:40:00	66.47
54	2/29/00	4:45:00	66.4
55	2/29/00	4:50:00	66.32
56	2/29/00	4:55:00	66.32
57	2/29/00	5:00:00	66.25
58	2/29/00	5:05:00	66.25
59	2/29/00	5:10:00	66.18
60	2/29/00	5:15:00	66.1
61	2/29/00	5:20:00	66.03
62	2/29/00	5:25:00	66.03
63	2/29/00	5:30:00	65.88
64	2/29/00	5:35:00	65.81
65	2/29/00	5:40:00	65.81
66	2/29/00	5:45:00	65.66
67	2/29/00	5:50:00	65.66
68	2/29/00	5:55:00	65.66
69	2/29/00	6:00:00	65.66
70	2/29/00	6:05:00	65.59
71	2/29/00	6:10:00	65.44
72	2/29/00	6:15:00	65.96
73	2/29/00	6:20:00	66.54
74	2/29/00	6:25:00	67.13
75	2/29/00	6:30:00	67.5
76	2/29/00	6:35:00	67.86
77	2/29/00	6:40:00	68.08

DataFeb2000

Figure 2.15

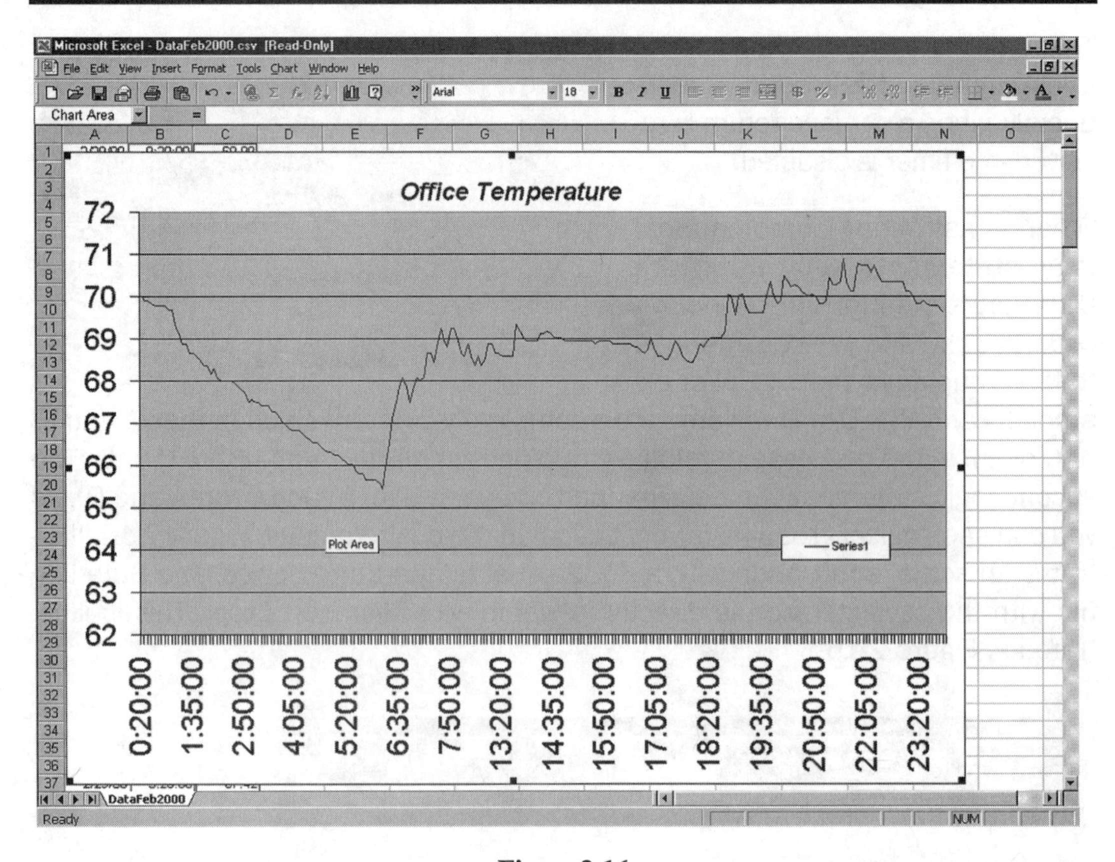

Figure 2.16

With the data loaded into Microsoft Excel, a data chart is just a few clicks away, as Figure 2.16 illustrates.

Conclusion/Exercise

This chapter introduced serial communication, provided some hands-on experience using the personal computer's serial port, and offered an opportunity to put some of the elements of Chapter 1 to work.

In the last section, General Purpose Serial Link, a general-purpose serial communications program was created. In addition to the serial communication exchange, this example also demonstrates how the data collected along with an associated time/date stamp is saved to a file for future reference. One item that was not discussed is error checking of the data. How do you know the data wasn't corrupted? Later chapters of this book will look into several types of error-checking schemes. There is a two-step, simple but effective error-checking scheme that can be added to the GPSL program.

As an exercise, you should implement the required code for the scheme thusly: Add another user-selectable field to the data-collection parameters. In this field, the user provides information on the expected response from the serial device. With the KT32 serial device, the response template would look something like "@+0000.00". The "+", "." and the "0s" indicate numerical characters and, therefore, represent the desired data field: a signed, six-digit number with two decimal places. All other characters are considered fixed literals. The received response would then be checked against this template. If an error occurred, another request would be generated. The second step of the error-checking scheme would require the data of two consecutive requests to be an identical numerical match. Similarly, a subsequent request would be issued until a match occurred.

Chapter 3
PLC FUNDAMENTALS

INTRODUCTION

This chapter provides general information about Programmable Logic Controllers (PLCs) and covers the following topics: a general description of a PLC, a Micro PLC memory map, the instruction set, and communication capabilities.

A PLC is microprocessor-based device specifically designed to interface with real-world devices through input/output modules. It consists typically of a Central Processing Unit (CPU), a proprietary operating system, nonvolatile memory, power supply, inputs ports, output ports, and communication ports. The CPU controls all the operations of the Programmable Logic Controller. Typically, the CPU is an Intel- or Motorola-class microprocessor.

Historically, the operating system is proprietary. Most of these operating systems were developed back when the cost of memory was at a premium and, as a result, the code is both lean and mean. Then again, the architecture only supports communications and the expansion of inputs and outputs. Furthermore, processing speed is very critical. It's crucial that the logical response to any input be as quick as possible.

Memory is provided for general-purpose use and for storing the user-provided program. The memory is nonvolatile, usually in the form of battery backed-up, low-power ram or Electrically Erasable Programmable Read Only Memory (EEPROM).

The power supply provided is extremely reliable and rugged. Conditioning is provided so that power-induced latch-up does not occur.

Conditioned inputs are provided for interfacing to real-world devices. The inputs are electrically isolated from the CPU power and the CPU data bus.

Outputs are provided with sufficient drive capability for activating lights and power relays. These outputs are typically in the form of relay contacts, open-emitter transistors, or field-effect transistors (FETs). As it is with the inputs, the outputs are electrically isolated from the CPU power and the CPU data bus.

Communications ports are provided for interfacing the PLC to programming devices, other serial devices, other PLCs, and other computers. These communication ports are used for programming and for exchanging information between other serial devices.

The PLC evolved from a need in the industrial world. Prior to PLCs, a desired combinational logic function for a particular control was achieved using hard-wired relay logic. Now, if the management of some industrial production facility decided to change the function (perhaps to increase production or to add a safety feature), the wiring between the various components would need to be modified. Furthermore, relays or timers may need to be added to implement the function. The additional components and cross wiring could result in a substantial amount of downtime for the production line and a substantial cost to implement.

The PLC revolutionized the industrial world by minimizing this process. With a PLC, input devices such as selector switches, limit switches, and proximity switches are wired directly to inputs of an input module. Output devices such as lights, horns, and motor-starters are wired directly to outputs of an output module. Physical devices such as timers, counters, sequencers, and closed-loop controllers are no longer required, since they can be created with software. The required function or functions are implemented and easily modified with software.

Ladder-Logic Programming

This software is specifically designed to emulate the relay ladder-logic diagrams that electricians had been working with for years. The most common form of programming a PLC is with ladder-logic programming, which is a visual form of programming. The left rail of the ladder represents the positive power; the right rail represents the common power point. Each rung of the ladder is an executable statement, with inputs originating on the left side and outputs terminating on the right side. Inputs are graphically represented by switching contacts or logical IF-type statements, while outputs are represented by relay coils or logical LET statements. If all the conditions are satisfied from the left rail up to the output, then that output or LET statement is activated. In addition, the software provides online monitoring where the ladder logic is animated by using colors to indicate the state of the various devices.

Figure 3.01 shows a single rung of ladder logic: If Input 1 is True and Input 2 is False then Output 1 becomes True. This statement has the same form as a Visual Basic statement. The only difference is that it's presented graphically with electrical symbols. An electrician would look at the rung of ladder logic as the conditions necessary to energize the coil. In this case, Input1 must be on (closed) and Input 2 must be off (open) for the coil to energize. At run time, the rung of ladder will be animated by the ladder-logic software by painting each device a particular color—for example, green for conducting and red for nonconducting. Just by looking at the ladder rung at run time, you can easily determine why a particular output or coil is not energized.

In addition to the memory allocated for ladder-logic storage, the PLC has memory available for general-purpose use. This memory is configured into 16-bit words called registers and is nonvolatile. Some PLCs configure memory as a large block of registers that can be accessed as a register or in bit form. Any one of these registers can be used in the construction of timers, counters, and other functions. Other PLCs have a memory system that is structured with a predefined allotment of registers for timers, counters, general-purpose storage, and bits.

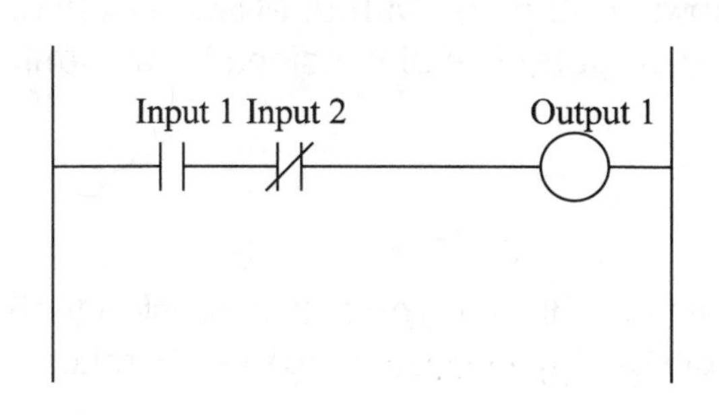

Figure 3.01

A conventional PLC consists of a rack that takes plug-in cards. Even the CPU is a plug-in card. This feature makes for a quick recovery of a downed system. There are many different types of input/output modules that can be installed in the rack. Some of these modules are: 24V DC input module, 120V AC input module, relay-output module, 120V AC triac-output module, analog input cards that take 4-20ma or 0-10 volts, analog output module that output 4-20ma or 0-10 volts, high-speed counter cards, Binary Coded Decimal modules, Servo modules, and so on.

Micro PLC

In recent years, a Micro PLC has been developed. As the name implies, Micro PLCs are physically small. The dimensions of one manufacturer's 10 Input/Output Micro PLC are 4.74" x 1.57" x 3.15". Typically, the Micro PLCs have a fixed amount of Inputs and Outputs. In addition to discrete inputs and outputs, some units also contain analog inputs and analog outputs.

All Micro PLCs provide communication. Most Micro PLCs are network capable. Some of the newer Micro PLCs today are becoming modular like their predecessor, the PLC. The Micro PLC contains a full-featured instructions set including both integer and floating-point math. Integer math deals only with whole numbers. Floating-point math allows you to use real numbers, for example, 1.25. The cost of a 10 Input/Output Micro PLC is approximately $200. This is a bargain when you consider the cost of purchasing discrete relays and timers—not to mention the time required to do the interconnection wiring.

Allen-Bradley is one of the largest manufacturers of PLCs in the world today. The MicroLogix 1000 is one family of Micro PLCs provided by Allen-Bradley. Although this chapter and subsequent chapters discuss the MicroLogix series of Micro PLCs, you will find that other Micro PLCs available from other manufacturers are very similar.

Figure 3.02 (courtesy of Allen-Bradley) shows a 32 Input/Output MicroLogix PLC. This model has 20 DC inputs and 12 Relay outputs. The dimensions for this unit are 7.87" x 3.15" x 2.87".

MicroLogix PLC Memory Map

In the MicroLogix PLC, each register consists of 16 bits. The first or least significant bit is bit 0. The most significant bit is bit 15. The memory map of a MicroLogix PLC is as follows.

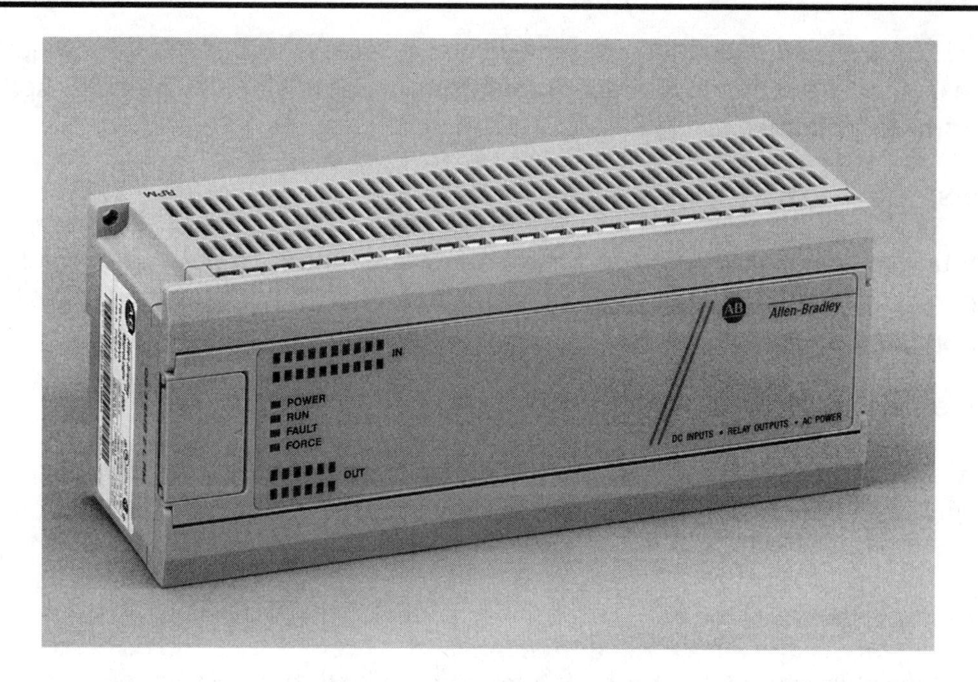

Figure 3.02

Outputs:

The number of outputs depends upon the model. The first output register is O:0.0. The first output, Output 0, is addressed as O:0.0/0. Output 11 is addressed as O:0.0/11.

Inputs:

The number of inputs depends upon the model. The first input register is I:0.0. The first input, Input 0, is addressed as I:0.0/0. Input 15 is addressed as I:0.0/15. Input 16 is addressed as I:0.1/0. Input 16 spills into the second input register I:0.1.

Internal coils:

Internal coils are classified as bits. 512 bits are allocated for internal coils. The bits start at address B3/0 and end at B3/511. They can also be addressed in register form starting at address B3:0 and ending at address B3:31. The bit form

B3/511 is the same location as the register address b3:31/15. In addition to these user bits, a number of predefined intrinsic status bits may be accessed for programming purposes.

Timers:

Forty timers are available in the MicroLogix PLC. Timers can be used to create time-on delay, time-off delay, or to keep track of time. The timers start at address T4:0 and end at address T4:39.

Counters:

Thirty-two counters are available for counting purposes. These counters can be configured to count up or down. The counters start at address C5:0 and end at address C5:31.

Control Registers:

Control registers store key information for specific instructions, such as shift registers and sequencers. There are 16 control registers available. The first control register's starting address is R6:0. The address of the last control register is R6:15.

Integer Registers:

There are 105 integer registers available for storing numerical values or bits. This memory is nonvolatile. The first integer register address is N7:0. The last integer register address is N7:104. Integer registers are bit-addressable. N7:0/0 is the least significant bit of register N7:0.

Before moving on to the MicroLogix instruction set, first examine how the PLC operates. PLCs work on a scan-based operating cycle. The operating cycle of most PLCs consists of the following steps.

1) Read the input table.
2) Execute the ladder logic.
3) Set the outputs according to the results of step 2.
4) Service any external communications.

5) Verify system reliability by performing a self-test. Go to halt mode on any failure. Typically, in halt mode, all outputs are disabled.

The scan time is a function of the size of the user program. Depending upon the size of the user program, the scan time can be 1 millisecond or 50 milliseconds. System reliability is a priority, and the PLC is always performing a self-test. You do not want a PLC operating machinery if the memory or any of the support chips should fail or become unpredictable.

MicroLogix Instruction Set

The instruction set for the MicroLogix is large, powerful, and diversified. This book devotes only one chapter to the PLC, so it's impossible to discuss each instruction. What follows is a mere sampling of the instruction set, focusing on the basic instructions.

Bit True (or Bit On) is graphically depicted in Figure 3.03. This symbol can reference any bit in the PLC memory. Bit False (or Bit Off) is graphically depicted in Figure 3.04. This symbol can reference any bit in the PLC memory.

Figure 3.03   Figure 3.04

Output Bit (or Coil) is graphically depicted in Figure 3.05. This symbol can reference any bit in the internal PLC memory and set it on or off based upon the conditions leading up to it. In addition, outputs can be specified as latched and unlatched. The symbols for latched and unlatched coils are shown in Figure 3.06. Once a latched coil is set on, it can only be cleared off by the unlatch instruction.

Figure 3.05   Figure 3.06

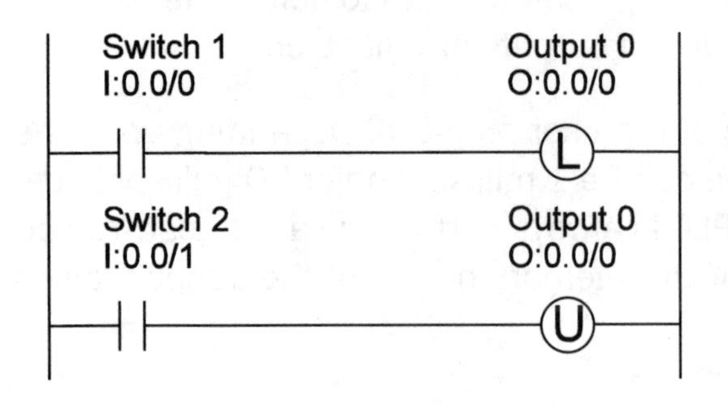

Figure 3.07

The ladder logic code illustrated in Figure 3.07 shows how the latched and unlatched coils can be used. In this example, both switch 1 wired to PLC Input 0 (or I:0.0/0) and switch 2 wired to PLC Input 1 (or I:0.0/1) are push-button switches. The latched unlatched output is external output 0 (or O:0.0/0). This output could be wired to turn on a light. When Switch 1 is depressed, the output turns on and stays on until switch 2 is depressed.

An output can also be instructed to go high for only one scan time by feeding the coil with a One-Shot Rising (OSR) instruction. The symbol for this instruction is depicted in Figure 3.08.

Figure 3.08

Figure 3.09 shows an example of ladder logic where, when switch one is closed, the internal coil N7:50/0 will go high for one scan time. The coil will only respond to an off-on or positive-edge transition. Internal coil B3/0 is utilized by the PLC operating system for the one-shot task. "Advance Counter" is a user-specified label for N7:50/0 that makes the ladder-logic software more readable. Labels are used to make the ladder-logic software readable.

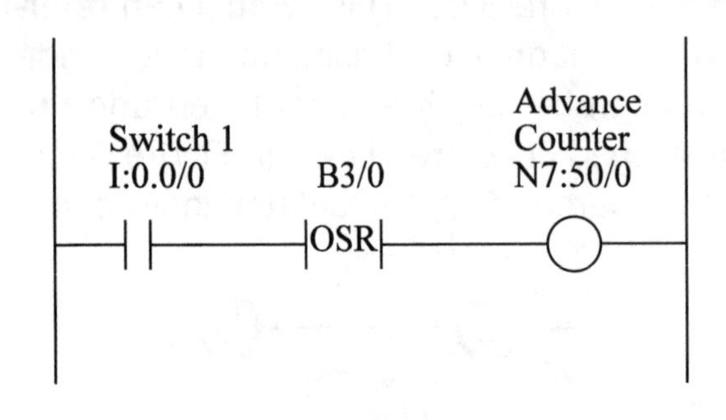

Figure 3.09

Three types of timers are available in the MicroLogix PLC. The first is the Timer on Delay (or TON), whose symbol is depicted in Figure 3.10. The MicroLogix memory supports 40 timers. Each timer is capable of counting up from 0 to a maximum of 32,767. The time base for each timer is selectable as 1 second or 10 milliseconds. Each timer has three status bits. The bit EN indicates that the timer is energized, or that all the elements feeding the timer from the opposite power rail are conducting. Bit TT indicates that the timer has not reached the specified preset and is still timing. Bit DN indicates that the timer has reached it preset value. In addition, the accumulator register, which the PLC operating system is using to keep track of the timer, is accessible to the ladder logic. Figure 3.11 provides an example of a timer application.

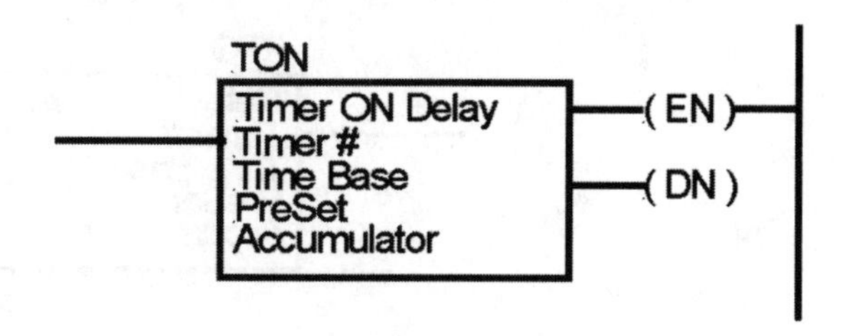

Figure 3.10

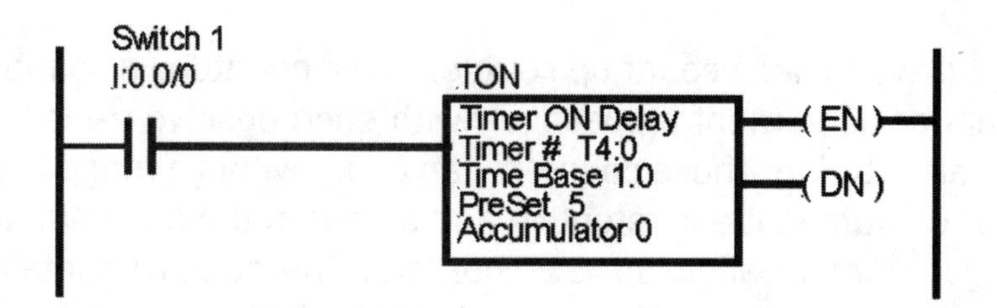

Figure 3.11

When Switch 1 is closed, the timer T4:0 begins counting in units of seconds and the accumulator register will advance with each passing second. The bit T4:0/EN and bit T4:0/TT are both energized (or True). If the preset value of 5 is reached, the accumulator stops advancing, bit T4:0/DN is energized (or True), and bit T4:0/TT is de-energized (or False). If Switch 1 is opened, the accumulator register clears and bits T4:0/EN and T4:0/TT are de-energized (or False). If bit T4:0/DN is energized before Switch 1 is opened, it will also be de-energized when Switch 1 is opened. The Timer-Off Delay timer is similar to the Timer-On Delay, except that it begins timing when the rung feeding it is de-energized. Figure 3.12 shows the symbol for the Timer-Off Delay timer.

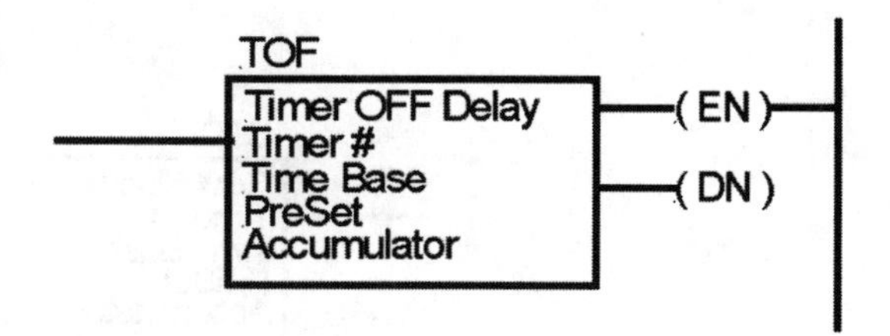

Figure 3.12

The third type of timer is called a Retentive Timer-On. This timer is identical to the Timer-on Delay, except that the value contained in the accumulator is not clear when the

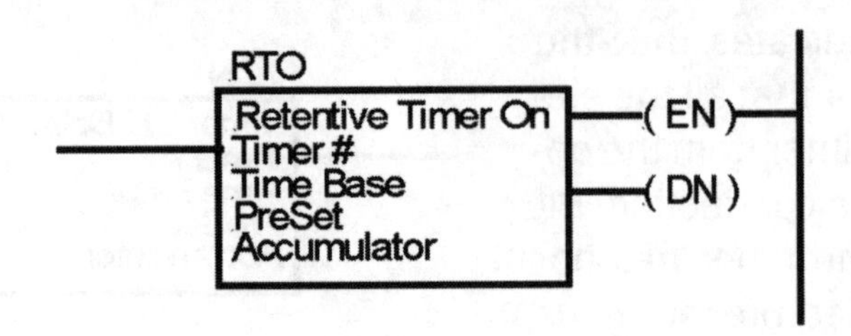

Figure 3.13

timer is de-energized. A separate instruction called a Reset must be asserted in order to clear the timer's accumulator. The symbol for the Retentive Time-On is shown in Figure 3.13.

Two types of counters are available: a count-up counter and a countdown counter. Each of these increment or decrement respectively with each positive transition or logical change from false to true. These counters can count within a range from -32,768 to +32,767. The count value is retentive and stored in the accumulator register. The MicroLogix memory allocates 32 counters. The counter contains status bits that indicate whether the counter is enabled, whether an underflow or overflow occurred, and when the value in the accumulator is greater than or equal to the preset value. Symbols for these counters are presented in Figure 3.14.

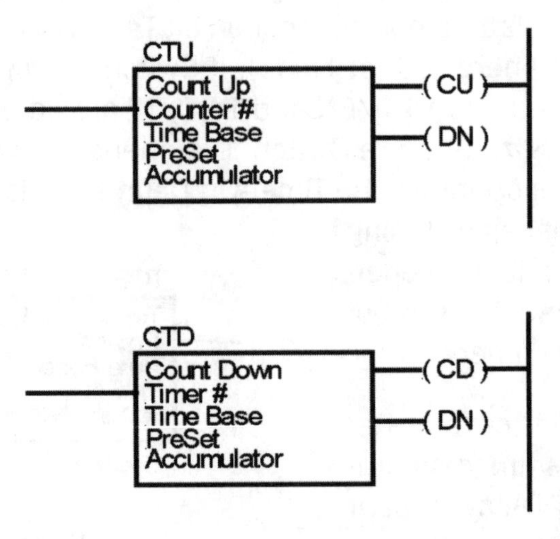

Figure 3.14

The following register-based data-handling instructions are available.

1) **TOD** Convert from integer to BCD. The maximum BCD number is 9999.
2) **FRD** Convert from BCD to integer.
3) **DCD** Decode a 4-bit value to single-bit value between 0 and 15.
4) **ENC** Encode a bit from 0 to 15 to a corresponding 4-bit value.
5) **COP** Copy a block of registers from one location to another.
6) **FLL** Fill a block of registers with a specified value.
7) **MOV** Move a register from one location to another.
8) **MVM** Move data from one location to mask specified part of an other location.
9) **AND** Perform a bit-wise logical AND Operation.
10) **OR** Perform a bit-wise logical OR Operation.
11) **XOR** Perform a bit-wise logical XOR (exclusive Or) Operation.
12) **NOT** Invert the state of each bit in a register.
13) **NEG** Change the sign of a register.
14) **FIFO** First-in/first-out operation.
15) **LIFO** Last-in/first-out operation.

The symbol for a Move instruction is shown in Figure 3.15. The Move instruction takes the contents of the register specified in the source location and places it in the register specified by the destination location.

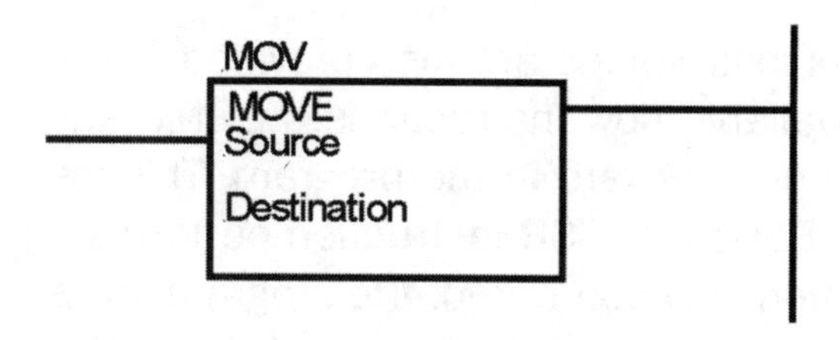

Figure 3.15

The symbol for an AND instruction is shown in Figure 3.16. This instruction takes the contents of Source A and logically ANDs the contents of Source B and places the result in the register location specified by the destination field. Source A and Source B can be either register locations or constant values.

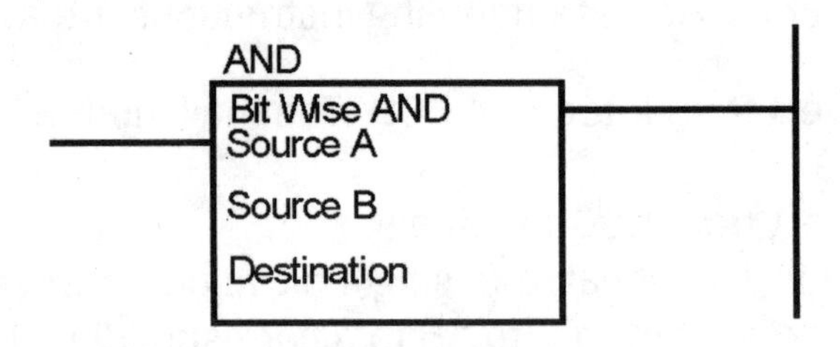

Figure 3.16

An example of an AND operation is presented in Figure 3.17. In this example, a value of nine is contained in Register N7:25. While Switch 1 is closed, the AND operation will be performed. The contents of N7:25, which initially contained 9, is now AND with the constant 7. The result of this operation is 1. Register N7:25 now contains a 1.

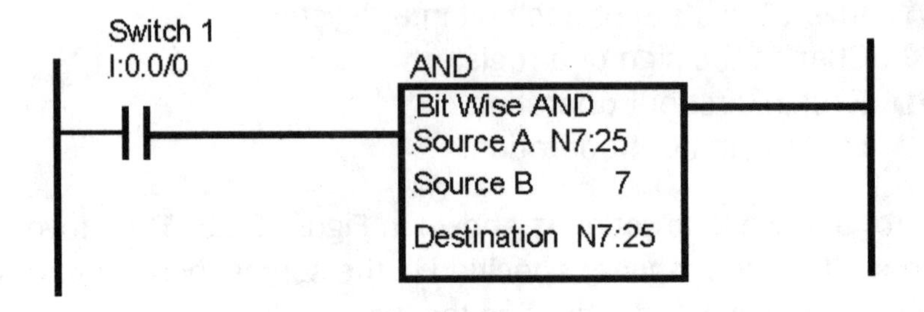

Figure 3.17

Program flow-control instructions are also provided in the Micro PLC instruction set. Some of the available flow instructions are JMP and JSR. JMP performs a jump either forward or backward in the program. This instruction is similar to a GOTO instruction in Basic. The JSR instruction performs a jump to a subroutine. When a JSR instruction is encountered, the program jumps to an area outside of the normal program area to execute code and then returns to the next rung of logic after the JSR.

The MicroLogix PLC is also capable of performing math instructions such as add, subtract, multiply, divide, and square root.

Lastly, the MicroLogix PLC has an instruction for messaging. With this, the PLC can initiate a communication message with another PLC or a PC that is directly connected to its serial communication port—or the PLC can relay the message to a network module, which then routes the request over the network. The message instruction provides for block moves of registers. The message can perform either a block read or a block write instruction.

Conclusion

Although the Micro PLC comes in a small package and with a small price tag, its power and capabilities are by no means small. Later chapters will examine in detail Visual Basic-based communications with the MicroLogix PLC.

Chapter 4
INTERFACING

INTRODUCTION

This chapter aims to familiarize you with the methods of interfacing to real-world devices. While large PLCs are rack-based and accept a variety of plug-in modules, most Micro PLCs have a fixed amount of discrete inputs and outputs that are of the same type. A 32 input/output (I/O) Micro PLC may consist of twenty 24V DC inputs and 12 relay outputs, or twenty 24V DC inputs and 12 FET (field-effect transistor) outputs. Most of the time, you will be required to interface to an input or an output that's not compatible with the existing I/O. All interfacing is based on the fundamental electrical principle, Ohm's law.

Ohm's Law

Georg Simon Ohm (1787-1854) was a German physicist. He was the first to formulate the relationship between voltage, resistance, and current. The Ohm, which represents a unit of electrical resistance, was named in his honor. The symbol used for the Ohm is the Greek letter *omega*, Ω.

Ohm's law states that voltage is directly proportional to the current multiplied by the resistance, or V (voltage)=I (current)*R (resistance). Since this is a mathematical function, you can derive the current to be the voltage divided by the resistance (I=V/R), or the resistance to be the voltage divided by the current (R=V/I).

This is a very simple principle that is perhaps difficult to grasp because you can't see voltage or current. Water pressure and water flow are analogous, respectively, to electrical potential (voltage) and electron flow (current). The water pipes in your house sit at a pressure level that is determined by either your water pump or a commercial water provider. The valve on your kitchen sink regulates the amount of water flow just as a resistor regulates the electron flow. As you close the valve, you increase the resistance of water flow, and the water flow drops off. Mathematically, I=V/R. As the magnitude of resistance increases, the magnitude of the current decreases proportionally—provided the voltage stays constant. Hopefully, this analogy helps you to comprehend Ohm's law. Ohm's law is at the very foundation of electrical theory, and if you do not understand it, then you are building your electrical knowledge on a poor foundation.

Now apply Ohm's law to limit the current through a light-emitting diode (LED) to 15 milliamps when attached to 24V DC supply. An LED is a polarized electrical component in which the anode is positive and the cathode is negative. The nominal turn-on voltage for an LED is 1.5V DC. This turn-on voltage varies based on the type of LED used. A manufacturer's specification will provide an exact range of values; however, a value of 1.5V DC is a good representative number to use for most LEDs.

In driving an LED, it's very important not to overdrive it with too much current and not to reverse bias it. LEDs are very polarity-sensitive; it only takes about 3 volts applied in the opposite direction to destroy them. If you were to apply a range of currents to a standard LED, you would find there isn't much change in brilliance from 14 milliamps to 20 milliamps (the typical maximum). A value of 15 milliamps is a good target biasing current for a standard LED. High-efficiency LEDs provides a good deal of light output at 1 milliamp. Figure 4.01 illustrates an LED circuit.

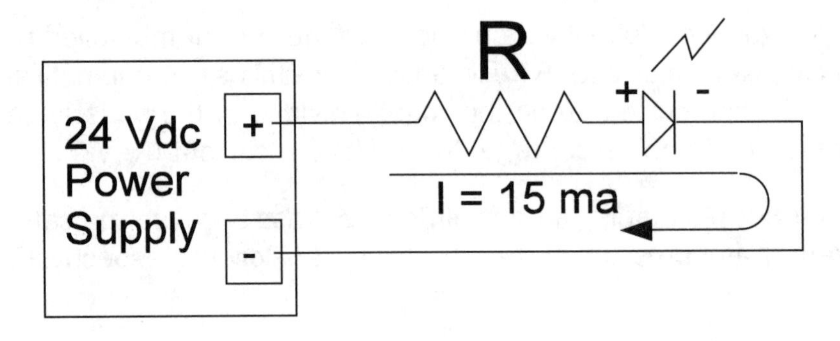

Figure 4.01

The voltage across the resistor R is equal to 24V DC minus the 1.5V DC voltage drop required by the LED device, or 22.5 V DC. Using Ohm's law, the resistance is determined by taking this voltage and dividing it by the desired current of 15ma. The required resistance is 22.5/.015 or 1500 ohms. 1500 ohms is a conventional resistance value. Another parameter that needs to be determined is the required power dissipation or wattage of the 1500 resistor. Power is equal to the voltage multiplied by the current (V*I). The power dissipation is approximately 22.5* .015, or .337 watts. Common wattage sizes for resistors are 1/8, 1/4, and 1/2. Therefore, a 1/2 watt 1500-ohm resistor is required for this circuit.

Interface Translators

One common form of voltage translator is called an optoisolator. This electronic device uses an infrared-emitting diode to couple photons to the base of an infrared photosensitive transistor. These electronic devices provide voltage translation and electrical isolation. In some cases, the gap between the emitter and detector is large enough to provide 7500 Vrms isolation. Most Micro PLCs' internal data and address buses are electrically isolated from the inputs and outputs by the use of optoisolators. The basic optoisolator is illustrated in Figure 4.02.

Electrically, the input to an optoisolator is configured like an LED. The forward voltage of the LED can be approximated to 1.5 volts; however, a new parameter is introduced called current-transfer ratio (CTR). The current-transfer ratio is the ratio of output-collector current Iout to the forward LED current Iin. This parameter is expressed as a percentage—usually as a minimum value. An optoisolator with CTR of 100% indicates a one-to-one relationship between LED current and transistor current. Optoisolators are available with a large variety of transfer

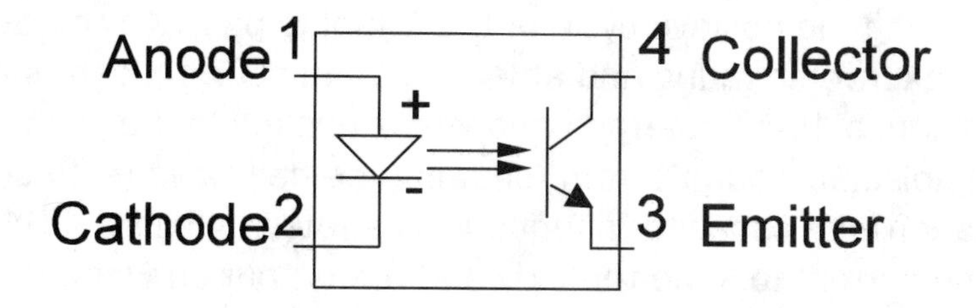

Figure 4.02

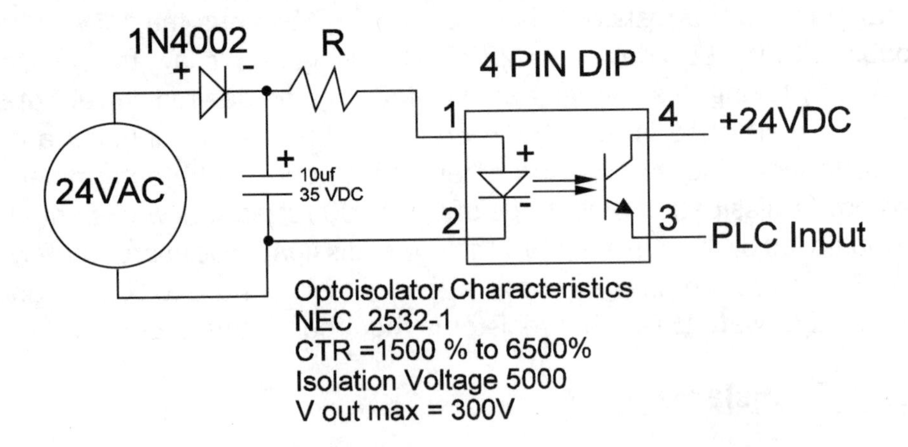

Figure 4.03

ratios, frequency responses, output voltages, output configurations, and output-current capacity. Here's how to use an optoisolator to interface to a Micro PLC 24V DC input.

The Allen-Bradley MicroLogix 24V DC inputs require a minimum of 2.5 milliamps at 14.4V DC to be considered in the "on" state. When 24V DC is applied, the input draws 8 milliamps. The circuit diagram in Figure 4.03 shows the requirements for interfacing to a Heating Ventilation Air Conditioning (HVAC) signal of 24V AC. This signal would be sent from a room thermostat to the HVAC unit instructing the unit to run.

The desired LED current in the optoisolator is 8 milliamps divided by the current-transfer ratio of 15, or .53 milliamps. The 24V AC supply used by the HVAC system is given terms of root-means-square (rms). The actual peak value of the sine wave is 1.414 times that amount, or approximately 34 volts. A diode is placed on the HVAC signal feed in order to keep the optoisolator voltage positive. The diode acts as a check valve, only allowing the positive cycle of the signal to pass. Any diode with at least a 100-volt peak inverse rating and at least a 1-amp current-carrying capacity will work. In addition, a 10-microfarad capacitor is applied to the circuit, in order to prevent the optoisolator output from following the half-wave rectified signal. Half-wave rectification with capacitor filtering is illustrated in Figure 4.04. The capacitor supplies the current to keep the LED and the output on when the

24 VAC Half Wave Retification with Capacitor Filtering

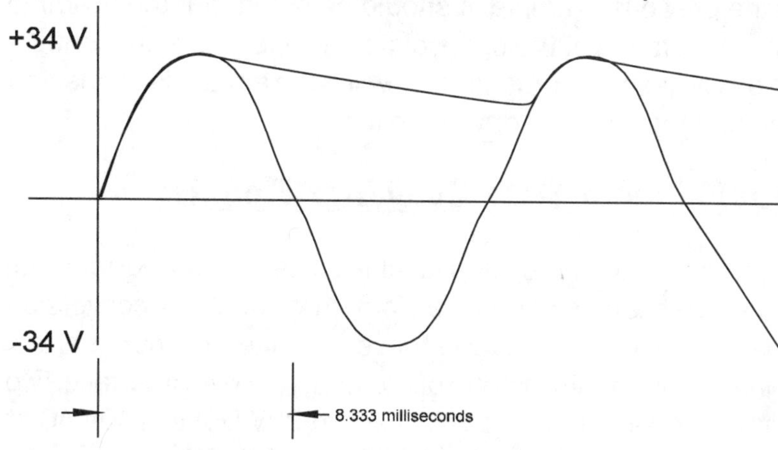

Figure 4.04

input signal swings negative. The average or DC value of the half-wave signal is V DC = .318 * V peak = 10.8 Volts. The required resistance is determined by 10 volts minus 1.5 volts (for the LED drop) divided by 0.53 milliamp and is equal to 16,038 ohms. The standard resistance value of 15 kohm should work just fine.

Another optoisolator example is presented in Figure 4.05. If you remember back in Chapter 2 with our first program, it was determined that the hardware handshaking signals could be used as a general-purpose computer controllable-output for some task. In Figure 4.05, an optoisolator is used to perform the electrical isolation and voltage translations tasks. The 1N4148 diode (or equivalent) prevents any negative voltage from damaging the optoisolator LED by clamping the voltage to diode drop.

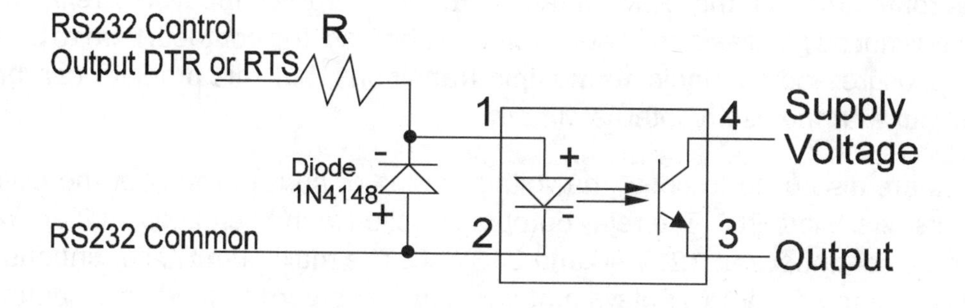

Figure 4.05

The value for the resistance can be determined based on output-current requirements, as presented in the previous example. It should be noted that the example depicts an open-emitter configuration of the optoisolator. An open-collector configuration could have just as easily been implemented, in which the emitter is connected to power supply common and the collector drives the load.

Commercially Available Solid-State Converter Modules

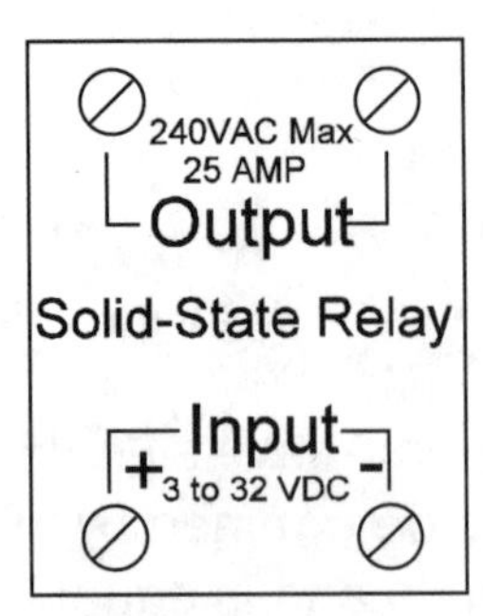

Figure 4.06

There are commercially available solid-state converter modules that can be used to perform just about any conversion task. You may want to use these packages for your interfacing requirements instead of creating your own interface.Two popular configurations are DC (3 to 32V DC) to 140V AC at up 50 amps and 120V AC to DC (3 to 60V DC).

These optically isolated, solid-state modules are four terminal devices—just as the example optoisolator is—except that the conditioning and packaging has been done at the factory. These converter modules can be found in most electronic catalogs. A solid-state relay module is illustrated in Figure 4.06.

Relays

Another common form of a voltage translator and electrical isolator is the relay. For example, if your Micro PLC has 24V DC inputs and you need to interface to a 120V AC input signal, a relay with a 120V AC coil can be utilized for the task. 24V DC is routed through the relay contacts into the Micro PLC input. The reliability of a relay is not as good as a solid-state relay. In addition, the contacts tend to bounce upon closure and can indicate multiple transitions, but this problem can be filtered out with the use of internal timers.

Relays are also used to boost the voltage rating or current rating of the existing outputs of a Micro PLC. The relay outputs supplied with MicroLogix PLC are rated 2.5 amp continuous at 120V AC and 240V AC. The maximum DC continuous current is 1 amp at 24V DC. Relays that can carry high current loads are commonly called contactors. Contactors with intrinsic motor-overload protection designed to operate motors are called motor starters. You could use a Micro PLC output to turn on or turn off a 450-horsepower, three-phase 460V AC induction motor by interfacing the output to either a reduced-voltage starter, a soft start, or a variable-frequency drive. These devices are capable of switching high-current motor loads and reducing the large inrush currents associated with starting motors. The reduced voltage starter accomplishes this task with contactors and a voltage-reducing transformer, whereas the soft-start and variable-frequency drive are electronically controlled, solid-state switching devices.

Analog-Based Sensors

Another interfacing consideration is the analog world. Temperature, pressure, acceleration, distance, flow, level, weight, and so on are quantitative physical entities. Transducers are used to convert these physical entities into corresponding voltages or currents. One popular voltage standard is 0 to 10 volts. A popular current standard is 4 to 20 milliamps. Micro PLCs with analog inputs convert these signals to an internal numeric value. This numeric value can then be used to turn on an output or implement some control function.

Now examine a pressure transducer that converts a pressure from 0 psi to 100 psi to a corresponding current source from 4 milliamps to 20 milliamps. Table 4.01 illustrates the current output for the 100 psi/ 4- to 20-milliamp pressure transducer at various pressures.

PSI	Current(milliamps)
0	4
25	8
50	12
75	16
100	20

Table 4.01

A current source is an electronic device that maintains a specified current. If you add resistance or remove resistance from a circuit being controlled by a current source, the current value will always be maintained. The electrical diagram in Figure 4.07 illustrates a circuit being driven by a 10-milliamp current source. Closing a switch shorts out the respective resistor. Ohm's law determines the voltage across Rout as 10 milliamps times 100 ohms, or 1 volt. If either switch S1 or switch S2 is closed, or if both are closed, the voltage across Rout remains at 1 volt and the currents stays at 10 milliamps. The 10-milliamp current is controlled by the current source. Any change in resistance—whether it's the resistance in the wire or corrosion at a terminal block–is automatically compensated for by the current source. The current source is a finite

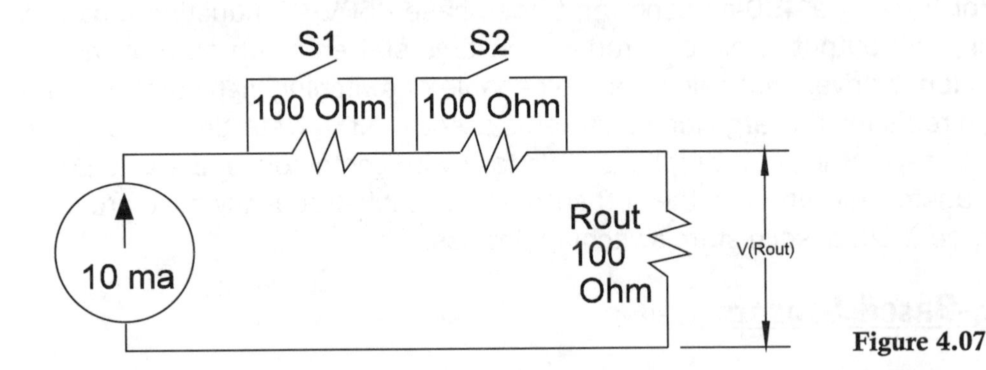

Figure 4.07

device that's only capable of driving a specified amount of resistance. If the Rout is removed from the circuit, or goes to a value of infinity, the voltage measured at Vrout is the maximum forcing voltage. Again, by using Ohm's law, the maximum load resistance can be determined by dividing the maximum forcing voltage by 20 milliamps. Another feature of a current source is that it's short-circuit proof; the current will remain at the specified value.

Most Micro PLCs that accept analog inputs accept 0 to 10 volts, 4 to 20 milliamps, or both. If only a 0- to 10-volt input was available, however, a transducer that outputs 4 to 20 milliamps could still be used by adding a 500-ohm resistor to convert the 4 to 20 milliamps to 2 to 10 volts.

Voltages greater than 10 volts can be monitored by using resistors to scale the voltage down to the 10-volt maximum. Figure 4.08 illustrates how to interface a 10-volt maximum analog input to monitor a 24-volt supply. The first resistor to determine is the 20-kohm resistor in Figure 4.08. You want this resistor to be large so that you are not pulling too much current from the supply. This is critical if you are monitoring a battery voltage. The resistance R is determined by subtracting the target voltage from the supply voltage and dividing by the current. In this case, the value of R is equal to 14/0.0005, or 28 kohms.

Analog-conversion modules capable of performing voltage or current translations on analog signals are commercially available. In addition, these conversion modules also provide electrical isolation. For example, a module can be purchased to

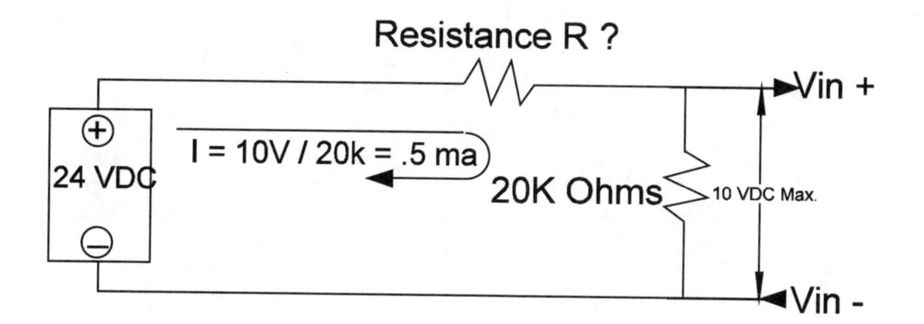

Figure 4.08

translate the 0- to 50-millivolt output of a load cell to a corresponding 4- to 20-milliamp signal. The 4- to 20-milliamp signal would then convey the load cell information into the Micro PLC.

Micro PLCs are also available with analog outputs. The popular, standard analog outputs are 0 volts to 10 volts and 4 milliamp to 20 milliamp. These analog outputs track according to a numeric value assigned by the Micro PLC. Typically, analog outputs are used for control application. Devices such as motor-speed controllers and proportioning valves have analog inputs capable of responding to remote commands.

Conclusion

Knowledge of Ohm's law, as well as familiarity with some of the commercially available conversion modules, make PLC interfacing an easy task. This chapter has provided a fundamental understanding of optoisolator components that can be utilized for any unique interfacing task.

Chapter 5
ALLEN-BRADLEY PROTOCOL

INTRODUCTION

PLC messaging consists of two types of messaging bytes: protocol bytes and data bytes. Protocol bytes provide the structure that encapsulates the data bytes, ensuring that the data bytes arrive undistorted at the correct destination.

Embedded in the firmware of the PLC is the required protocol for messaging. In order to communicate with this device, both the protocol and the memory structure of the PLC must be understood. The PLC memory structure was covered previously and is somewhat similar from one PLC manufacturer to another. Protocol, on the other hand, is a whole new ball game.

Typically, the original equipment manufacturer (OEM) will provide supplemental reference manuals that provide intrinsic details about their product. These reference manuals allow third-party developers to develop products that ultimately enhance the OEM's product line. This type of reference manual will need to be purchased.

The Protocol

Allen-Bradley provides a reference manual entitled "Data Highway/Data Highway Plus/ DH-485 Communication Protocol and Command Set," which describes the protocol required to talk to their family of PLCs. This reference manual is available for purchase. The publication number that this book is working from is 1770-6.5.16, dated November 1991. The Allen-Bradley Part number is 404610904.

Rockwell Automation, which owns Allen-Bradley, was kind enough to grant this author the permission to write about two commands. These two commands are an Unprotected Read and an Unprotected Write. With these two commands, your objective to exchange data with a PLC can be achieved. Again, it should be noted that Rockwell Automation/Allen-Bradley in no way implies or states any certification or approval of the material discussed in this book.

The Unprotected Read command and the Unprotected Write command permit access to only the PLC's Data memory. The maximum amount of contiguous memory that can be read with a read command is 122 words. Similarly, a write command permits 121 contiguous words to be written with a single write command. The Allen-Bradley byte transmission format is as follows.

start bit—8 bit data—no parity—stop bit

The framework of the data packet is constructed with ASCII control characters. In the ASCII chart, the control characters reside from 0 to 1F in hexadecimal (0 to 31 decimal), and at the very end of the chart is "DEL" at 7F hex or 127 decimal.

Actually, the ASCII control characters can be further grouped into physical device controls (linefeed, carriage return, etc.), logical communication controls (i.e., SOH meaning start of header), physical communication controls (such as NUL), and controls for code extensions. The original creators of ASCII wanted to cover all bases. Supplemental information on the ASCII character set should be easy to locate.

It's unfortunate that, although an ANSII standard was instituted that provides a graphical representation of ASCII control characters, this standard is not readily available. If one looks up the ASCII character set in Visual Basic Help, all the control characters are graphically represented as □ —not much help if one is streaming characters into a textbox for troubleshooting purposes.

The Allen-Bradley protocol uses several of the control characters that are located in the logical communication control group. Refer to the following list.

ASCII Abbreviation	Hexadecimal Value	Decimal Value
STX	02	2
ETX	03	3
ENQ	05	5
ACK	06	6
DLE	10	16
NAK	15	21

The classical definition of each character is as follows.

STX "Start of Text" marks the beginning of data often referenced in ASCII as text.
ETX "End of Text" marks the end of data often referenced in ASCII as text.
ENQ "Enquiry" is issued when a response is expected but not received.
ACK "Positive Acknowledge."
DLE "Data Link Escape" is used to indicate a control sequence.
NAK "Negative Acknowledge."

Allen-Bradley uses the term symbol, which is defined as two or more bytes having a specific meaning in the protocol. The following list depicts the symbols required for communications.

1) **DLE STX** This symbol indicates the start of a message.
2) **APP DATA** This symbol represents a series of bytes that represent the data.
3) **DLE ETX BCC/CRC** This symbol represents the end of the message. BCC (Block Character Check) and CRC (Cyclic Redundancy Check) are error-checking routines that ensure the quality of the message. These routines will be examined later in this chapter.
4) **DLE ACK** This symbol indicates a positive confirmation.
5) **DLE NAK** This symbol indicates a negative response to a message.
6) **DLE ENQ** This is a sender symbol that requests a retransmission of a response from the receiver.
7) **DLE DLE** Any time that a data value 16 (DLE) exists in the APP DATA section, it must be transmitted as a double byte DLE-DLE. This indicates

that it is data, rather than a control character. For example, the data section may contain a sequence of data bytes 16 followed by 2, which could easily be interpreted as a DLE STX or a start of message. A sequence of 3, 16, 16, 2 in the APP DATA component indicates a true data value of 3,16, 2.

Allen-Bradley Communication Packet Structure

An Allen-Bradley communication packet is depicted in the following figure. This figure applies specifically to the unprotected read and unprotected write commands. Other commands will have a format that is specific to the type of command. These commands are listed in the Allen-Bradley literature.

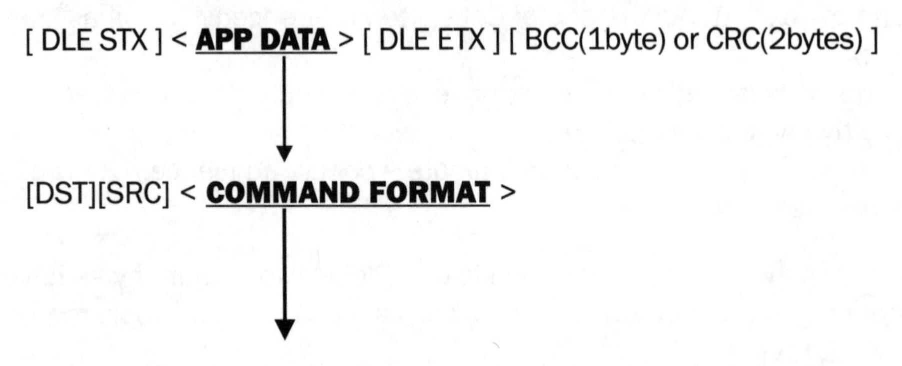

[CMD][STS][TNS1][TNS2][ADDL][ADDH]<DATA (write)/ SIZE (read) >

DST is the message-destination node address.
SRC is the source of the message node address.
CMD is the command byte (Unprotected read=1) (Unprotected write=8).
STS is a status byte that typically equals zero.
TNS1 is the low byte of a unique 16-bit transaction identifier sent with each request.
TNS2 is the high byte of a unique 16-bit transaction identifier sent with each request.
ADDL is the low order address byte.
ADDH is the high order address byte.
DATA specifies the data to be written. PLC words are two bytes long.

SIZE specifies the number of words to be read. PLC words are two bytes long.

Addressing is specified because network modules are optionally available for this family of Micro PLCs. With these network modules, the PLCs can be linked together and exchange data between each other by intrinsic messaging commands, as well as with a personal computer or personal computers.

Communication Packet Error Checking

The communications protocol specifies two types of message-wide error-checking techniques: Block Check Character (BCC) and Cyclic Redundancy Check (CRC). In addition, a lower level error check can be performed at the byte level by enabling parity checking.

Studies have shown, however, that parity checking is only capable of providing minimal error detection. The communication message byte (Start—8bit data—Stop) would result in a 10% increase in time for minimal-at-best error detection. This unproductive increase in time should be avoided. Let Parity equal “NONE”.

Block Check Character (BCC)

The Block Check Character (BCC) error-checking algorithm provides a medium level of detection and is easy to implement. The Allen-Bradley BCC requires a two’s complement sum of all the bytes between DLE STX and DLE ETX. Note that if any of the bytes are equal to 10 Hex thus DLE, then two DLE must be transmitted but only one is used in the BCC calculation. Essentially, all the APP LAYER bytes are arithmetically summed together. Then the result of the summation is exclusive Or’d with FF Hex. Finally, a 1 is added. The least significant byte of the result is the BCC.

In order to clarify the BCC concept, here’s an example. All the values are in hexadecimal.

The actual APP LAYER = FF|12|A0|00|03|10|7   Note the DLE(10 Hex)

The transmitted packet is as follows.

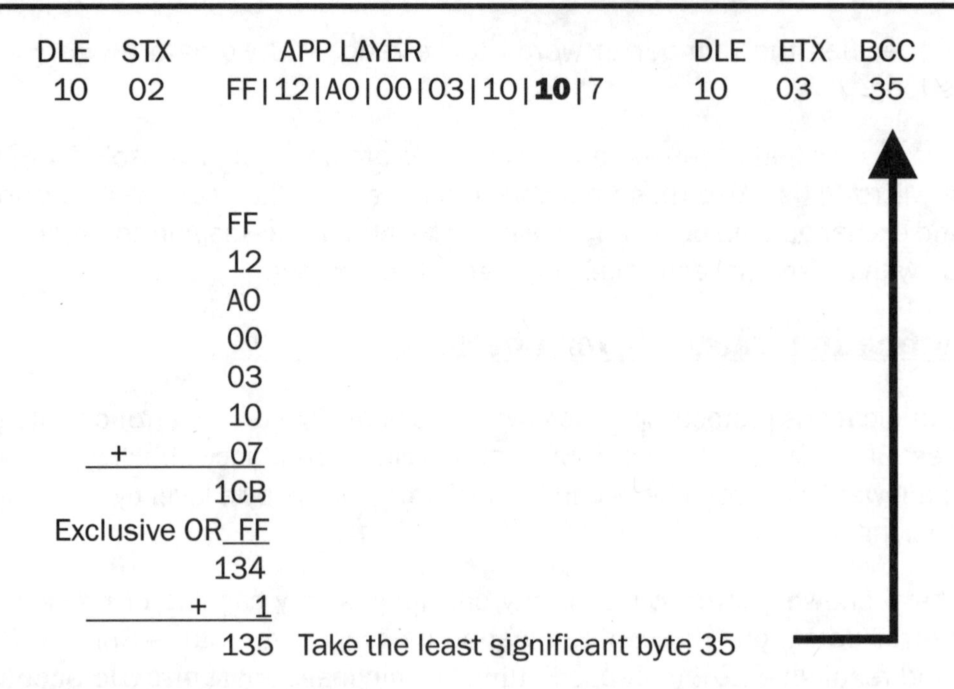

The shortcomings of BCC are somewhat apparent. If any zero bytes were to be inserted by noise, or if the bytes were to be transposed, the BCC would be correct but the APP LAYER would be in error.

Cyclic Redundancy Check (CRC)

Cyclic Redundancy Check provides an extremely high level of error detection and is more difficult to implement. The theory behind the CRC is beyond the scope of this book; however, there are many communication books that discuss this theory in detail. What's important is that you know how the PLC computes the CRC. The following description details Allen-Bradley CRC generation. The protocol specifies CRC-16, which essentially means that the result will consist of 16 bits. Two bytes are required to be transmitted for this type of error check. The least significant byte is transmitted first.

Like the BCC, the CRC operates on all the bytes of the APP LAYER. In addition, an ETX or 03 Hex is included in the calculation. The Allen-Bradley CRC function is depicted in the following algorithm.

```
Let DataByte = APP LAYER + ETX
Let CRCinteger = 0
Let CRCconstant = A001 Hex (integer dimension)

For Each DataByte
  Let CRCinteger = CRCinteger Exclusive OR with the DataByte
        (This operates only on the least significant byte of CRCinteger)
  Do Eight(8) Times
    Shift CRCinteger to the right; shift a zero into the most significant bit
      If the bit shifted out of the least significant bit is a 1 then
        Let CRCinteger = CRCinteger Exclusive OR with the CRCconstant
      End If
  End Do
End For

CRCinteger contains CRC
```

Needless to say, you wouldn't want to calculate out a multiple-byte CRC by hand. From a programming perspective, this function is relatively easy to implement. The Shift right function is binary division, or equivalent to dividing by two. Care must be taken when implementing this function that rounding from the division does not distort the result.

Unprotected Write Command Structure

This command will write data into PLC memory. The message packet is constructed as follows.

[DLE STX][DST][SRC] < **COMMAND FORMAT** > [DLE ETX] [BCC or CRC]

DST is the message destination node address.
SRC is the source of the message node address.
CMD is the command byte.
STS is a status byte that typically equals zero.
TNS1 is the low byte of a unique 16-bit transaction identifier sent with each request.

TNS2 is the high byte of a unique 16-bit transaction identifier sent with each request.
ADDL is the low order address byte.
ADDH is the high order address byte.

Command Format:
CMD = 8
[CMD][STS][TNS1][TNS2][ADDL][ADDH][DATA – maximum of 244 bytes]

Reply Format:
CMD = 72 = 48 Hex
[CMD][STS][TNS1][TNS2]

TNS1 and TNS2 are bytes that contain a transaction value. The protocol specifies that each command issued must have a transaction number that is different than the last command issued. If the transaction number does not differ, the unit won't respond. This doesn't mean that you have to implement a 16-bit number. The communication would work, if the transaction number simply alternated between zero and one with each command.

But it's not that difficult to set up a 16-bit counter that increments with each transmitted command. The Reply Format transaction number is the same number that was contained in the Command Format.

In the Allen-Bradley MicroLogix PLC, an address of zero (ADDL=0, ADDH=0) corresponds to PLC memory location for integer file type identified as N7. N7 is a non-volatile area of PLC memory that has a capacity of 104 16-bit registers.

The following command (all numbers shown in hexadecimal) will write an 8001 Hex into the contents of register N7:0. Remember that a PLC register is 16 bits, or 2-bytes wide.

[CMD][STS][TNS1][TNS2][ADDL][ADDH][DATA – maximum of 244 bytes]
[8][0][34][12][0][0][01] [80]

The Reply message for this command is as follows.

[CMD][STS][TNS1][TNS2]

[48][0][34][12]

Note that the transaction number transmitted from the PLC in response to the Unprotected Write command is identical to the transaction number sent with the command. This characteristic further ensures the transmitter that the receiver responded to its command.

Unprotected Read Command Structure

This command reads data from the PLC memory. Like the Unprotected Write command, the message packet is constructed as follows.

[DLE STX][DST][SRC] < **COMMAND FORMAT** > [DLE ETX] [BCC or CRC]

Command Format:
CMD = 1
[CMD][STS][TNS1][TNS2][ADDL][ADDH][SIZE]

Reply Format:
CMD = 65 = 41 Hex
[CMD][STS][TNS1][TNS2][DATA – maximum of 244 bytes]

The size field specifies the amount of bytes to be read. A single PLC memory location consists of two bytes. The low order byte is always transmitted first.

In the Allen-Bradley MicroLogix PLC, an address of zero (ADDL=0, ADDH=0) corresponds to PLC memory location for integer file type identified as N7. The following command (all numbers shown in hexadecimal) will read three data bytes starting at register N7:1.

[CMD][STS][TNS1][TNS2][ADDL][ADDH][SIZE]
[1][0][AB][CD][2][0][3]

The Reply message for this command is as follows.

[CMD][STS][TNS1][TNS2][DATA]
[41][0][AB][CD][11][22][33]

The PLC responded to the Data request sending the required data. Like the Unprotected Write Command, the transmitted transaction bytes are echoed back with the response. Both the size field and the address field are specified quantitatively in bytes. The low address field contains a 2, which points to a PLC address of N7:1 and does not point to PLC address N7:2. Each PLC memory location consists of two bytes. If you wanted to read the contents of N7:50, a value of 100 would need to placed in the low address field.

The transmitted data reveals that the contents of PLC memory location N7:1 is 2211 Hex and that the least significant byte of PLC memory location N7:2 is 33 Hex. Please note that this information was empirically determined with a Visual Basic program developed later in this chapter.

Command/Response Exchange

In addition to the message-packet structure discussed, the Allen-Bradley protocol specifies a sequence of responses for the exchange. This sequence is as follows.

Command Sender

1) Send a command and initialize timers.
2) Wait for a DLE ACK.
3) If a DLE NAK is received, then retransmit the command. Do this up to X amount of times.
4) If neither a DLE ACK nor a DLE NAK was received within a specified time period, then transmit out a DLE ENQ (enquiry). Do this up to X amount of times.
5) If a time-out occurs, then proceed to transmit out the next request.

Command Receiver

1) If a message packet is received, then verify the integrity of the message by checking the BCC or CRC.
2) If the integrity of the message is bad, then transmit out a DLE NAK and wait for the next message.
3) If the integrity is good, then transmit out a DLE ACK. Proceed to generate the response.
4) Send the response and initialize timers.
5) Wait for a DLE ACK.

6) If a DLE NAK is received, then retransmit the command. Do this up to X amount of times.
7) If neither a DLE ACK nor a DLE NAK was received within a specified time period, then transmit out a DLE ENQ (enquiry). Do this up to X amount of times.
8) If a time-out occurs, then proceed to wait for the next request.

The value for "X" can be specified in the MicroLogix Micro PLC. The default value for MicroLogix series "C" and later is 6.

First PLC Data-Acquisition Program

As it turns out, the requirements for proper communication to the Allen-Bradley MicroLogix PLC are pretty straightforward and don't appear to be very difficult to implement with Visual Basic code. The PLC is capable of serial communication at standard Baud rates from 300 to 38400, with 9600 being the default value specified by the factory. All of these Baud rates are within the realm of the Visual Basic 6 Communication control.

Since data acquisition is very critical to our desired application of a home monitor, it's recommended that a simple program be developed to test the waters, so to speak. You need to be able to write data to the PLC and read data from the PLC. So tackle this task and create a simple program that sets the outputs of the PLC and reads the PLC inputs. A virtual PLC will be created on a Visual Basic form. The input and output status lights of this virtual PLC will mirror the status lights of the real PLC. Activating an input on the real PLC will light the respective input light on the virtual PLC. Similarly, by clicking on the output status light of the virtual PLC, the virtual light and corresponding real output change state.

Although you could communicate directly with the input and output data files of the PLC, it's more efficient to transfer the PLC I/O to a PLC data area with a PLC move instruction. This permits the inputs to be conditioned with timers for debouncing or some other type of conditioning. In addition, there may be other internal registers, which are some function of the inputs that you want to read.

By moving this data so that it's contiguous, you need only supply a single read command to fetch a large amount of data. This process is commonly called packing.

Similarly, with a write command, you may want to drive an output in addition to permitting the PLC to drive this output with logic of its own accord. By writing to a data area, this combinational logic is possible by programming the PLC. In addition, there may be other areas of memory that need to be written. Again, by structuring the write data so that it's contiguous, you need only supply a single write command to efficiently transmit a large block of data.

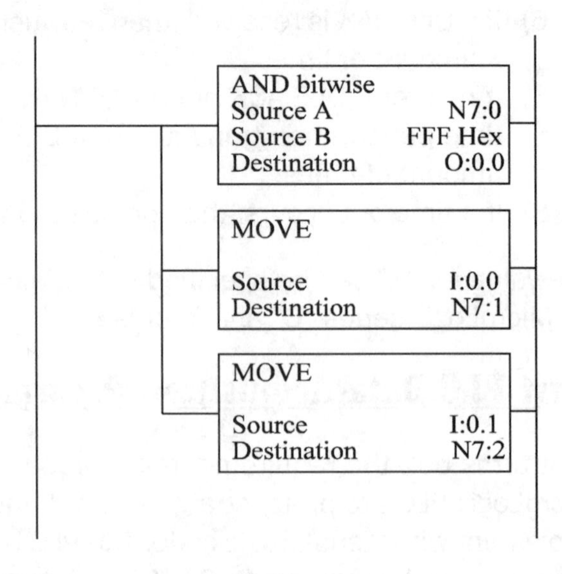

Figure 5.01

This first PLC data-acquisition program is going to communicate with an Allen-Bradley 32 I/O MicroLogix PLC. This PLC has (20) 24V DC inputs and 12 relay outputs. PLC memory location N7:0 will be used to drive the relay outputs.

Since this PLC has 12 outputs, four bits of the 16-bit register N7:0 are left over. PLC memory location N7: 1 and N7: 2 will contain the status of the PLC inputs. The first 16 inputs consume all the bits of register N7: 1. This is not the case with register N7: 2, in which the remaining inputs use four out of 16 bits. These spare inputs and outputs could be used to manipulate internal functions.

PLC Ladder Logic Requirements

Since the acquisition program is targeting a data area of PLC memory, the following single rung of ladder logic code provided in Figure 5.01 will transfer the data appropriately from data area to the PLC inputs and outputs.

The first instruction executes a logical AND operation on the data contained in register N7:0 with a constant FFF Hex and places the result in the PLC output file. The AND function is used to ensure that a PLC fault is not generated by inadvertently trying to turn on a nonexistent relay. This PLC only has 12 outputs. If bits 12 through 15 were to be high, the PLC would go into a fault mode because these outputs don't exist. The subsequent MOVE instructions transfer the data from the PLC input files to Data-area registers N7:1 and N7:2. If you are a puzzled over the PLC memory nomenclature, you might want to review Chapter 3.

Form Design

Since the function of our first PLC communication program is to transport the PLC inputs to the computer and transport the computer outputs to the PLC, the title of the form is "MicroLogix Data Transporter." Figures 5.02 and 5.03 show the layout of the form. All the coding for this program is located on the companion CD-ROM in the folder entitled "Chapter 5."

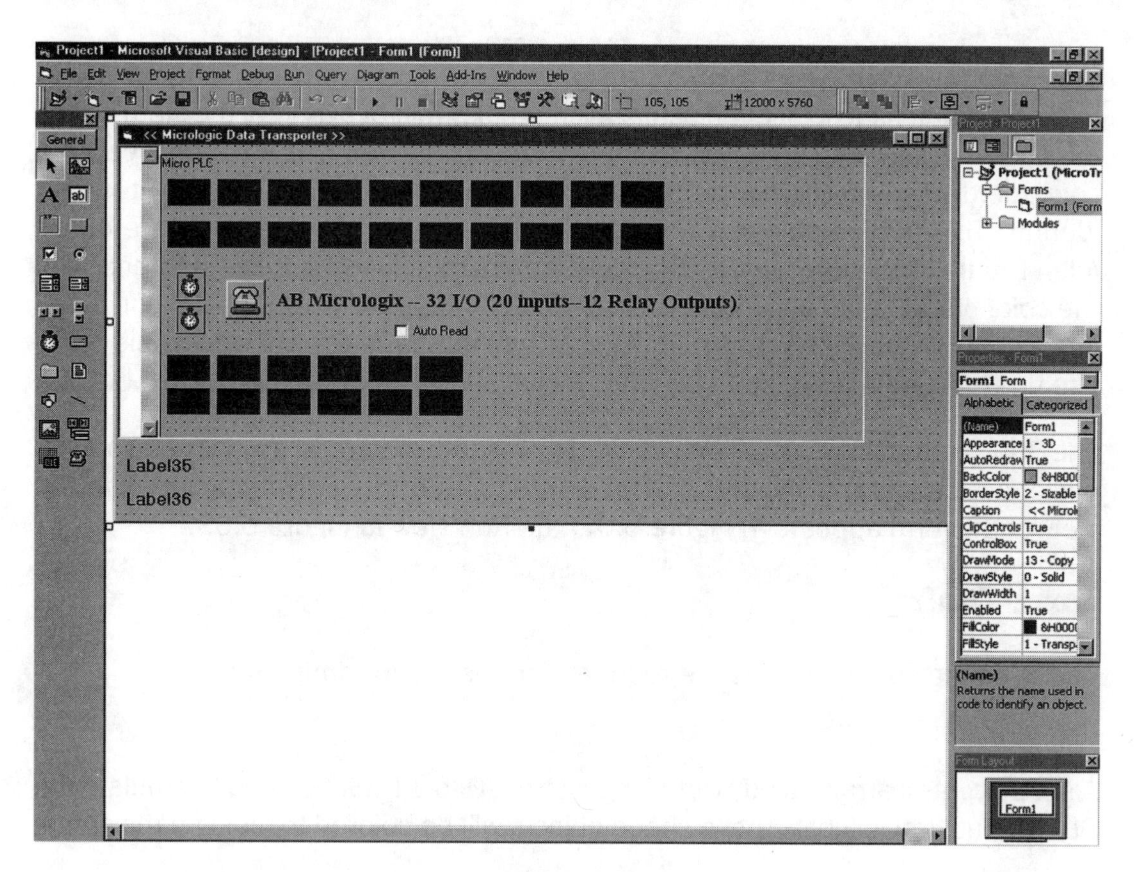

Figure 5.02

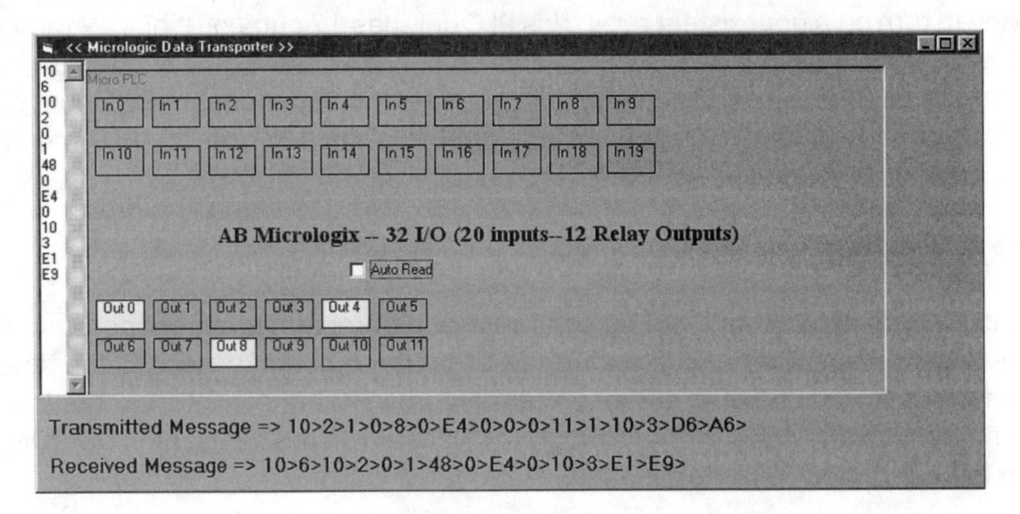

Figure 5.03

In order to match the look of an actual Allen-Bradley MicroLogix PLC, the status lights are constructed with a shape object and an associated label object. The labels are configured with a transparent background. The label caption is the status-light identity. The status lights are arranged in multiple rows just as they are on an Allen-Bradley MicroLogix PLC. You could have simply used a Label and changed the color of the Label background accordingly—much like what was done in previous chapters—but then you wouldn't be using the Shape control. You would really use this technique to animate circular or oval shapes that required a text overlay.

The Shape control is a visual object that is available in several predefined shapes: circle, oval, square, rectangle, rounded square, and rounded rectangle. The IDE view of the form appears in Figure 5.02; the Run view in Figure 5.03.

Form Controls

The following control groups are required for this programming task.

PLC outputs:

A shape-control array called "Out" ranging from 0 to 11 will be used to emulate the PLC output-status lights. These shape objects will be colored to indicate the status of the output light. Filling the shape with yellow will indicate an ON state. Filling the shape with gray will indicate an OFF state. The array approach will help to streamline and minimize the coding of this form.

A control array is created in the following fashion. First, a shape object with the desired characteristic is created. In this case, we want a rectangular shape for the light with a solid fill style. The desired control name is changed to “Out.” Once the size of the shape has been determined, the shape object is selected with a left click of the mouse. Then, with a right-click of the mouse, select copy. Under the Edit menu, select paste to create a copy on the form. A dialogue box will appear prompting as to whether a control array should be created. Select Yes. The copied control is automatically placed in the upper left-hand corner of the form. Move the control to the desired location. Paste this control on the form until a total of 12 shapes have been created.

Each output-shape object has an associated Label control that overlays the shape object. The Label-caption property is used to identify the underlying associated shape with text, such as “Out 1”. The BackStyle property of the Label is selected as transparent so that the Label background does not interfere with coloring of the shape. In addition, the Label- event procedure “Click” for mouse click will be utilized to change the state of the output-status light and to generate the required Unprotected Write command to the PLC. The size of the Label should be identical to the underlying shape, since the Label is used to determine the mouse-click event.

Once the Label has been sized and configured accordingly, it too can be copied and pasted on the form. In this instance, a control array is not desired. The program provided on the companion CD-ROM has label1 through label8 associated with Out(0) to Out(7), respectively. Label31 through label34 are associated with Out(8) through Out(11), respectively.

It should be noted that this approach was taken to provide an example using a combination of a shape control with an associated Label. The technique is ideal if the shape requirement is round or oval. This animation could have been accomplished by simply coloring the background of each Label.

PLC inputs:

The PLC input-status lights will be emulated with a shape-control array called "Zin" ranging from 0 to 19. These shape objects will be colored to indicate the status of the output light. Filling the shape with yellow will indicate an ON state. Filling the shape with gray will indicate an OFF state. The array approach will streamline and minimize the coding of this form. Again, the BackStyle property of the Label is selected as transparent so that the Label background does not interfere with coloring of the shape.

Like the PLC output-status lights, Label controls will be used to provide a textual identifier for each input-status light. The naming of these Label controls is not critical, since they only provide text information.

General-Purpose Controls:

An MSComm control provides the serial interface for the project. This ActiveX control is not contained in the standard toolbox. It's added at design time by right-clicking on the toolbar and selecting "components". Under the control tab, use the vertical scroll bar to find and place a check mark on "Microsoft Comm Control 6.0".

Two timers provide communication-timing functions. One timer is used to create a polling function for the PLC inputs and the other provides a communications time-out.

A textbox named "RX" is used to capture and display all of the received bytes. This textbox control is provided in the event that a large message is transmitted. The vertical scroll bar will assist in the inspection of the received bytes. This control may not be necessary for this project; however, since you're exploring an existing communication system, it may come in handy to provide insight into the Allen-Bradley protocol.

A checkbox control is provided in order to stop the automatic polling function. This freeze-frame feature will facilitate inspection of the contents of both the transmitted and received bytes.

Lastly, various Label controls provide miscellaneous functions. One Label is used to group the PLC inputs and outputs into a virtual PLC. Another Label indicates the PLC type and the amount of inputs and outputs. Label35 is used to show the transmitted bytes in hexadecimal delimited with a ">". Change the name of Label35 to lblTxtData. Label36 is used to show the received bytes in hexadecimal delimited with a ">". Change the name of Label36 to lblRxData.

The Code

You should start by defining key variables and placing them in the general section of a module.

```
Option Explicit

'COMMUNICATION VARIABLES AND CONSTANTS

Public Const STX As Byte = 2
Public Const ETX As Byte = 3
Public Const DLE As Byte = 16
Public Const ACK As Byte = 6
Public Const ENQ As Byte = 5
Public Const NAK As Byte = 21
Public Const STS As Byte = 0
Public bDST As Byte
Public bSRC As Byte
Public bCMD As Byte
Public bLCRC As Byte
Public bHCRC As Byte
Public bRxLCRC As Byte
Public bRxHCRC As Byte

Public bDATA As Byte     ' first byte of write command
                         ' twelve outputs require 1.5 bytes
Public bDATA2 As Byte    ' second byte of write command

Public bLastLastRX As Byte   'last last receive byte
Public bLastRX As Byte       'last receive byte
Public bNewRX As Byte        'newest receive byte
Public bRXpointer As Byte
Public bETXposition As Byte
```

```
Public bReadData As Byte        'first byte of the read command
                                '20 inputs requires 2.5 bytes
Public bReadData2 As Byte       ' second byte of the read command
Public bReadData3 As Byte       'third byte of the read command

Public STXflag As Boolean
Public ETXflag As Boolean

Public WriteModeFlag As Boolean
Public ReadModeFlag As Boolean
Public bComStep As Byte
Public Const lCRCconstant As Long = 40961 ' HEX A001
Public lTransaction As Long

Public bTxData(100) As Byte
Public bRxData(100) As Byte
```

You may be wondering why a long data type is specified for both the CRC constant and the transaction counter. While both of these may be 16-bit values, Visual Basic defines integer variables as two-byte numbers ranging in value from -32,768 to 32,767. Any time the most significant bit (sign bit) is set to a 1, an overflow error will be generated. Dimensioning these variables as Long eliminates this overflow error.

CRC Sub Procedure

Earlier in this chapter, you determined the requirements of a CRC error-checking scheme. Now, implement the code for this task as a Sub procedure in a module.

```
Public Sub COMPUTECRC()

Dim CRCREGISTER As Long 'this is just a scratchpad variable specific to this sub
Dim t As Byte
t = bTxData(0) + 1 'VB6 always allocates array(0) because the default lower
                   'bound is zero unless an "Option Base 1" statement is placed in
                   'the declaration section of a module. Let's use this byte
                   to point to the end of the data in the array.
```

```
 bTxData(t) = ETX 'LOAD IN AN ETX AS REQUIRED FOR CRC CALCULA-
TION AFTER THE LAST DATA BYTE IN THE ARRAY.
Dim D As Byte
 Dim SHIFTFLAG As Boolean

 CRCREGISTER = 0

 For t = 1 To (bTxData(0) + 1) 'FOR ALL DATA INCLUDING THE ADDED ETX
 CRCREGISTER = Int(CRCREGISTER Xor bTxData(t))
       For D = 1 To 8
        'If the least significant bit is high this is equivalent to shift out so per-
form
        'an Exclusive Or of the CrcRegister with the CRCconstant
      If (CRCREGISTER And 1)=1 Then SHIFTFLAG=True Else SHIFTFLAG= False
       CRCREGISTER = Int(CRCREGISTER / 2) 'The Int keyword removes
the fractional part of the 'number and returns the resulting integer value.
This prevents rounding which distorts the result.
    If SHIFTFLAG=True Then CRCREGISTER=CRCREGISTER Xor lCRCconstant
    Next D
Next t

bLCRC = CRCREGISTER And 255
bHCRC = Int(CRCREGISTER / 256) ' Use the Int keyword to prevent rounding

End Sub
```

Sub Form Load() Event Code

The Form_Load Event is the region where the project is initialized. An inspection of the following code will reveal the advantage of using a control array for the input and output shapes. With a few lines of code, all 32 shape objects are initialized to the OFF state.

```
Private Sub Form_Load()

Dim T As Byte  'use T as a scratch pad byte
For T = 0 To 19
```

```
'Since there are only twelve outputs while T is < 12 color fill the outputs
 If T < 12 Then out(T).FillColor = QBColor(7)
 zin(T).FillColor = QBColor(7)
Next T

' Initialize auto read as off

Check1.Value = 0  'show checkbox as unchecked
READtimer.Enabled = False 'turn auto read timer off

' Set up the Communications Port
  MSComm1.CommPort = 2  ' change according to your system configura-
tion
   ' 9600 baud, no parity, 8 data, and 1 stop bit.
   MSComm1.Settings = "9600,N,8,1"
   ' Tell the control to read entire buffer when Input is used.
   MSComm1.InputLen = 1
  MSComm1.RThreshold = 1 'Set the characters to return to 1
   ' Open the port.
   MSComm1.PortOpen = True
  'set up PLC Communication Parameters
   bDST = 1 'destination address: The PLC default is 1.
   bSRC = 0 'source address:  the computer is address zero
  READtimer.Interval = 50 ' Every 50 Millisecond the timer fires
  READtimer.Enabled = True 'Start the read timer
  ComTimeout.Enabled = False 'Make sure the ComTimeout timer is dis-
abled.
End Sub
```

Sub Form Unload() Event Code

Use the Form_Unload event to make sure that the communications port is turned off.

```
Private Sub Form_Unload(Cancel As Integer)

'Turn off the communication port
```

```
MSComm1.PortOpen = False
End Sub
```

Animation of the Output-Status Lights

In this project, the output-status lights of the actual PLC are to track according to the lights of the computer-based representation of the PLC. If you click on an output-status light with the mouse, it is to change state accordingly, and an Unprotected Write command is issued to the actual PLC. If the PLC is a model type with relay outputs, you should instantaneously hear the relay click in or click out as the status light is toggled with each click of the mouse. When a PLC output is on, the status light turns yellow. This PLC has 12 outputs; therefore, two bytes are required to be transmitted to the PLC.

The following Click-event Sub for label1 illustrates the required code. The QBColor keyword is used to fill the shape with gray (QBColor7) if it is off and yellow (QBColor14) if it is on. The Tag property of the shape is, conveniently, used to store a numerical value for each output. When the Label is clicked, if the FillColor of the shape is yellow, then the FillColor of the shape is changed to gray and the Tag value of the shape is changed to zero. If the FillColor of the shape is gray, however, then the FillColor of the shape is changed to yellow and the Tag value of the shape is changed to a number. The number placed in the Tag value of the shape is based upon a binary progression. Label1 is linked to Out(0), so the on value is 2 raised to zero, or 1. For label8, which is linked to Out(7), the value is 2 raised to seven, or 128. When the output is off, the Tag value is set to zero for all cases. Out(8) through Out(11) values are 1, 2, 4, 8 of the second write byte.

```
Private Sub Label1_Click()
    If out(0).FillColor = QBColor(14) Then
            out(0).FillColor = QBColor(7): out(0).Tag = 0
    Else: out(0).FillColor = QBColor(14): out(0).Tag = 1
    End If
Plcwrite 'A sub procedure
End Sub
```

•

•

•

```
Private Sub Label8_Click()
    If out(7).FillColor = QBColor(14) Then
            out(7).FillColor = QBColor(7): out(7).Tag = 0
    Else: out(7).FillColor = QBColor(14): out(7).Tag = 128
    End If
Plcwrite 'A sub procedure
End Sub


Private Sub Label31_Click()
    If out(8).FillColor = QBColor(14) Then
            out(8).FillColor = QBColor(7): out(8).Tag = 0
    Else: out(8).FillColor = QBColor(14): out(8).Tag = 1
    End If
Plcwrite 'A sub procedure
End Sub
```

•

•

•

```
Private Sub Label34_Click()
    If out(11).FillColor = QBColor(14) Then
            out(11).FillColor = QBColor(7): out(11).Tag = 0
    Else: out(11).FillColor = QBColor(14): out(11).Tag = 8
    End If
Plcwrite 'A sub procedure
End Sub

Private Sub plcwrite()
'This Sub procedure sums all the shape tags together into two bytes for
PLC transmission
```

```
'DATA = Val(out(0).Tag) + Val(out(1).Tag) + Val(out(2).Tag) + Val(out(3).Tag)
        +Val(out(4).Tag) _
        + Val(out(5).Tag) + Val(out(6).Tag) + Val(out(7).Tag)
Dim T As Byte
bDATA = 0
For T = 0 To 7
  bDATA = bDATA + Val(out(T).Tag) ' The first byte of write data
Next T
'the second byte of write data
bDATA2=Val(out(8).Tag)+Val(out(9).Tag)+Val(out(10).Tag)+ Val(out(11).Tag)
WriteModeFlag=True 'initiates a write request to the program by setting
the flag

End Sub
```

Now that the PLC outputs have been assimilated into two bytes, a flag called WriteModeFlag is set high, indicating that a write command should generated. The method that generates the write command will be discussed later in the chapter. Ultimately, the transmit procedure sends these bytes to the PLC register N7:0. The PLC ladder logic then routes the contents of register N7:0 directly to the PLC outputs.

Animation of the Input-Status Lights

This task requires that the PLC inputs be polled at some periodic rate. The method that generates the read command will be discussed later in the chapter. Basically, at a rate determined by the Readtimer, an Unprotected Read command solicits the PLC to read three bytes of data starting at register location N7:1. The ladder logic in the PLC moves the input states into registers N7:1 and N7:2 on every PLC scan. Three bytes are required to be collected because this PLC has 20 inputs. These bytes are placed in ReadData, ReadData2, and ReadData3. The following Sub procedure called ShowInputStatus illustrates how these bytes are dissected and used to animate each respective input-status light. Again, indexing is used to perform this task. As T increments from zero though seven, each byte is simultaneously ANDed with the respective bit pattern (01,02,04...128). If a nonzero is detected, the input is on and the respective shape is filled with yellow. Otherwise, the respective shape is filled with gray, indicating it is off.

```
Private Sub ShowInputStatus()
    Dim T As Byte
 For T = 0 To 7
'animate inputs 0 thru 7
    If (bReadData And (2 ^ T)) <> 0 Then
            zin(T).FillColor = QBColor(14) 'Yellow
      Else: zin(T).FillColor = QBColor(7) 'Gray
    End If
'animate inputs 8 thru 15
    If (bReadData2 And (2 ^ T)) <> 0 Then
            zin(T + 8).FillColor = QBColor(14) 'Yellow
     Else: zin(T + 8).FillColor = QBColor(7) 'Gray
    End If
'animate inputs 16 thru 19
    If T < 4 Then  'Do only for the first 4 bits
     If (bReadData3 And (2 ^ T)) <> 0 Then
            zin(T + 16).FillColor = QBColor(14) 'Yellow
     Else: zin(T + 16).FillColor = QBColor(7) 'Gray
     End If
    End If
Next T
End Sub
```

Communication Procedural Aspects

Moving right along to the communication procedural aspects of this project—essentially, it's a very simple task. Fortunately, the PC is the one doing the polling, which makes the communication part easier.

The primary function is to exchange information between the PC and the PLC. The PC is aware when the output information changes because of the output-status lights-label click event. When an output change is detected, the output information is transmitted to the PLC. In order to effectively monitor the PLC inputs, the PC needs to poll the PLC as often as possible. At 9600 bps and 10 bits per character, each byte time interval is approximately 1.04 milliseconds. The longest message transaction is the read request. The minimum number of bytes for the read command—including request, response, and the corresponding Acks—is 34 bytes. The total transaction time is 35.4 milliseconds. This doesn't include any multiple DLEs or 10's hexadecimal that may be contained in the data or in the transaction number. The readtimer interval can be safely set to 50 milliseconds with no chance of overlap. Therefore, the computer will receive update information from the PLC 20 times per second.

Every 50 milliseconds a message transaction will occur. Normally this transaction will be a read request, unless one of the PLC status lights on the Visual Basic form is clicked on and the WriteModeFlag is set, or the "Auto Read" checkbox is unchecked. The Auto Read checkbox was placed on the form so that the message structure can be examined. The following code illustrates the message-request flow.

```
Private Sub READtimer_Timer()
If bComStep = 0 Then
   'This path if a message transaction is not going on
   'Increment the transaction number
 If lTransaction >= 65535 Then
  lTransaction = 0
  Else: lTransaction = lTransaction + 1
 End If

  If WriteModeFlag = True Then
   GOWRITE
   ComTimeout.Enabled = True
   bComStep = 1
 Else:
  If Check1.Value = 1 Then
    ReadModeFlag = True
    GoRead
    ComTimeout.Enabled = True
    bComStep = 1
  End If
 End If
End If
End Sub
```

The ReadTimer event initiates all the communications events, as depicted in the previous code. The communication procedural process can be represented by five steps. The code tracks these communication steps with the variable bComStep. A description for each communication step follows.

Communication Step 1 (bComStep=1): Send the request and wait for acknowledgment from the PLC. Enable the ComTimeOut timer. If a NAK is received, then retransmit the message and continue to wait. If an ACK is received, move on to Step 2.

Communication Step 2 (bComStep=2): Collect the response. Begin looking for a DLE STX. If a DLE STX is detected, then start collecting the APP layer and strip off any redundant data DLEs. DLEs are 10 hexadecimal. Start looking for a DLE ETX, which indicates the end of data. Once a DLE ETX is detected, the next two received bytes represent the received CRC. Compare the calculated CRC to the received CRC and move on to Step 3.

Communication Step 3 (bComStep=3): Generate an acknowledgment. If the calculated CRC and the received CRC are equal, then transmit out an ACK, disable the ComTimeOut timer, and move on to Step 4. If the CRCs do not match, then transmit out a NAK, reset all receive flags, and remain at Step 2.

Communication Step 4 (bComStep=4): Process the Data. Do supplemental testing of the data packet by checking for the proper size, proper destination, and proper transaction number before processing the data. Move on to Step 5.

Communication Step 5 (bComStep=0): Communication transaction is complete. If a DLE ENQ is received, then retransmit an ACK. The PLC did not receive the ACK transmitted in Step 3. Step 5 can be verified by commenting out the TxAck sub procedure call in Step 3. You will then see that the PLC will generate a DLE ENQ (10>5). These values will be appended to the receive string.

If the reception was a response to a write command, then that's the end of the messaging. If the received message was a response to a read request, however, then the received bytes are routed into the read data bytes, the "ShowInputStatus" Sub Procedure is executed, and the status lights of the computer PLC mirror the actual PLC.

If a communication time-out occurs, then the messaging attempt is abandoned, all flags are reset, and the next message is sent. A time-out is normally caused by a disconnection in the communication cable, or by powering off the PLC.

The following code satisfies the remaining requirements of the project.

```
Private Sub GOWRITE()
  Dim G As Byte 'G is a scratch pad byte that serves as a counter of
            ' bytes entered into the transmit array
  bCMD = 8 'IN Allen-Bradley Protocol an Unprotected Write command is=to 8.
  G = 1
  bTxData(G) = bDST  'Destination address
  G = G + 1
   bTxData(G) = bSRC 'Source address
  G = G + 1
  bTxData(G) = bCMD 'WRITE
  G = G + 1
  bTxData(G) = STS ' Status Byte Constant
  G = G + 1
  bTxData(G) = (lTransaction And 255)
  'ITRANSACTION is incremented before each new read or write request
  ' when ITRANSACTION exceeds FFFF HEX it is cleared to 0000
  G = G + 1
  bTxData(G) = Int(lTransaction / 256) 'use Int to prevent rounding
  G = G + 1
  bTxData(G) = 0 ' write address low byte N7:0
  G = G + 1
  bTxData(G) = 0 ' write address high byte
  G = G + 1
  bTxData(G) = bDATA  'low byte  represents outputs 0 through 7
  G = G + 1
  bTxData(G) = bDATA2
  bTxData(0) = G  'Place the total count in array location zero
 ClearAllcomflags
  COMPUTECRC ' computes the crc based on data contained in bTxData array.
  '  bLCRC is low byte of the result. bHCRC is the high byte of the result.
```

```
    GOTRANSMIT ' Sends the message out of the Communication Port.
End Sub
```

```
Private Sub GoRead()
  ' Load up the Read array
   Dim G As Byte
   bCMD = 1
   G = 1
   bTxData(G) = bDST
   G = G + 1
   bTxData(G) = bSRC
   G = G + 1
   bTxData(G) = bCMD 'Read command
   G = G + 1
   bTxData(G) = STS
   G = G + 1
   'lTransaction is incremented before each new read or write request
   bTxData(G) = (lTransaction And 255)
   G = G + 1
   bTxData(G) = Int(lTransaction / 256)
   G = G + 1
   bTxData(G) = 2
   G = G + 1
   bTxData(G) = 0
   G = G + 1
   bTxData(G) = 3  'size
   'place the end of data pointer in the array zero location
    bTxData(0) = G
 ClearAllcomflags
   COMPUTECRC ' computes the crc based on data contained in bTXDATA array.
   '  LCRC is low byte of the result. HCRC is the high byte of the result.
   GOTRANSMIT
 End Sub
```

```
 Public Sub GOTRANSMIT()
```

```
txack 'Transmit out an ACK just in case there are any unanswered requests
      ' These can occur by starting and stopping the program.
lblTxtData.Caption = "Transmitted Message => " 'initialize transmit read-
out
lblRxData.Caption = "Received Message => " 'initialize receive readout
bLastLastRX = 0 'initialize receive history for ACK/NAK response
bLastRX = 0
bNewRx = 0
 ' Start of Messaging Packet
MSComm1.Output = Chr(DLE)
'Chr() Returns a String containing the character associated
'with the specified character code
lblTxtData.Caption = lblTxtData.Caption + CStr(Hex(DLE)) + ">"
'tack each transmitted character on to label caption with a ">" delimiter
MSComm1.Output = Chr(STX)
lblTxtData.Caption = lblTxtData.Caption + CStr(Hex(STX)) + ">"

Dim TXTpointer As Byte
 ' Start of APP Layer (All DLE's should be doubled up)
For TXTpointer = 1 To bTxData(0) ' bTxData(0) contains size of APP Layer

MSComm1.Output = Chr(bTxData(TXTpointer))
lblTxtData.Caption = lblTxtData.Caption + CStr(Hex(bTxData(TXTpointer)))
+ ">"
 'If the transmitted byte is a DLE, transmitted a second DLE
 If bTxData(TXTpointer) = DLE Then
   MSComm1.Output = Chr(DLE)
   lblTxtData.Caption = lblTxtData.Caption + CStr(Hex(DLE)) + ">"
 End If
Next TXTpointer

' End of Messaging Packet
MSComm1.Output = Chr(DLE)
lblTxtData.Caption = lblTxtData.Caption + CStr(Hex(DLE)) + ">"
MSComm1.Output = Chr(ETX)
lblTxtData.Caption = lblTxtData.Caption + CStr(Hex(ETX)) + ">"
MSComm1.Output = Chr(bLCRC)
```

```
lblTxtData.Caption = lblTxtData.Caption + CStr(Hex(bLCRC)) + ">"
MSComm1.Output = Chr(bHCRC)
lblTxtData.Caption = lblTxtData.Caption + CStr(Hex(bHCRC)) + ">"

'Initialize the time-out timer
ComTimeout.Enabled = True
Dim s1 'Clear receive buffer
s1 = MSComm1.Input

End Sub
```

```
Private Sub MSComm1_OnComm()
'Static s1 As String
Dim s1 As String
Select Case MSComm1.CommEvent

Case comEvReceive   ' Received RThreshold # of chars.
   s1 = MSComm1.Input
   RX.Text = RX.Text + CStr(Hex(Asc(s1))) + vbCrLf
   lblRxData.Caption = lblRxData.Caption + CStr(Hex(Asc(s1))) + ">"

   bLastLastRX = bLastRX 'save last three responses
   bLastRX = bNewRx
    bNewRx = Asc(s1)

    If bComStep = 2 Then
       If STXflag = True Then
         GATHERresponse
       Else: If bLastLastRX<>DLE And bLastRX=DLE And bNewRx = STX Then _
       STXflag = True: bRXpointer = 1
       End If
   End If

 End Select

  If bComStep = 1 Then 'Step 1 wait for Ack or Nak after issuing request
  If STXflag = False And bLastRX = DLE And bNewRx = NAK Then GOTRANSMIT
```

```
If STXflag = False And bLastRX = DLE And bNewRx = ACK Then
  bComStep = bComStep + 1 'Step 2: The uPLC received the request
   'Clear receive history and wait for DLE STX
  bLastLastRX = 0
  bLastRX = 0
  bNewRx = 0
  End If
End If

If bComStep = 0 And bLastRX = DLE And bNewRx = ENQ Then
txack 'Resend Ack the uPlc did not receive it at step 4
bLastLastRX = 0
bLastRX = 0
bNewRx = 0
End If

End Sub
```

```
Private Sub GATHERresponse()
 Dim DLEstripFlag As Boolean
bRxData(bRXpointer) = bNewRx
 'strip off any extra DLE's
If bRXpointer > 1 And ETXflag=False And bNewRx=DLE And bLastRX=DLE Then_
bRXpointer = bRXpointer - 1: DLEstripFlag = True: bNewRx = 0: bLastRX = 0

'the App layer starts with  [DST][SRC][CMD][STS][TNS1][TNS2]
' bRXpointer settings          1    2    3    4    5     6
'no point in looking  for an ETX until the pointer is > 6
If bRXpointer > 6 And ETXflag = False And bNewRx = ETX _
 And bRxData(bRXpointer - 1)=DLE And DLEstripFlag=False Then ETXflag=True:_
bETXposition = bRXpointer

'Determine if the CRC bytes have been collected
If ETXflag = True And bRXpointer = bETXposition + 2 Then
bRxData(0) = bETXposition - 2
bRxHCRC = bRxData(bRXpointer)
bRxLCRC = bRxData(bRXpointer - 1)
```

```
Dim t As Byte
    For t = 0 To bRxData(0)
    bTxData(t) = bRxData(t)
    Next t
   COMPUTECRC

  If bRxHCRC <> bHCRC Or bRxLCRC <> bLCRC Then
   txNAK
   ClearAllcomflags
End If

If bRxHCRC = bHCRC And bRxLCRC = bLCRC Then 'Received Good Message
 bComStep=bComStep + 1 'Step 3 Received Good Response Send Ack to uPLC
  ComTimeout.Enabled = False 'Stop the Time out Timer
  txack 'Acknowledge the response
  bComStep = bComStep + 1 'Step 4  Process Data

  bComStep = 0 'Clear Step to Zero the request is complete
   If ReadModeFlag = True Then 'collect the data
    ' move to a normalized array
    If (bETXposition - 6) = 5 Then
    If bRxData(1)=0 And bRxData(5)=(lTransaction And 255) AndbRxData(6)=_
         Int(lTransaction / 256) Then

         ReadModeFlag = False
          bReadData = bRxData(7)
            bReadData2 = bRxData(8)
              bReadData3 = bRxData(9)
                ShowInputStatus
       End If
     End If
   Else: WriteModeFlag = False
   End If
 End If
End If

bRXpointer = bRXpointer + 1
```

```
End Sub
```

```
Private Sub txack()
MSComm1.Output = Chr(DLE)
MSComm1.Output = Chr(ACK)
End Sub
```

```
Private Sub txNAK()
MSComm1.Output = Chr(DLE)
MSComm1.Output = Chr(NAK)
End Sub
```

```
Private Sub ClearAllcomflags()
bLastLastRX = 0
bLastRX = 0
bNewRx = 0
STXflag = False
ETXflag = False
bRXpointer = 1
bETXposition = 0
 RX.Text = ""
End Sub
```

Conclusion/Exercise

Given the material presented earlier in the chapter, and with the descriptions provided in the code itself, you should be able to understand and follow the logic of the code. If you are still having trouble, then you should go back and review the start of this chapter and follow the code step by step. Similar code will be discussed again in the next chapter as you set out to create the home-monitoring acquisition program.

As an exercise, you should modify the code so that the PC outputs always reflect the true state of the PLC's actual outputs. The program, as it stands, will load with all the outputs in the off state by default. It's not until the first write command that the PC outputs match the PLC outputs.

Chapter 6
HOME-MONITOR PROJECT

INTRODUCTION

The objective of this chapter is to define the scope of the home-monitor project, to create an interconnection diagram between the required components and the Allen-Bradley MicroLogix PLC, and to develop the required ladder logic software.

PLC Task & Interconnection Diagram

As a home monitor, the PLC will monitor the following digital inputs.

1) Front doorbell push button
2) Rear doorbell push button
3) Front door sensor open/closed status
4) Rear door sensor open/closed status
5) First floor HVAC system on/off status
6) Second floor HVAC system on/off status
7) Water pump on/off status
8) Mailbox sensor open/closed status

Besides the discrete digital inputs, the following analog inputs will be monitored.

1) Outside air temperature
2) Crawl-space temperature
3) First floor temperature
4) Second floor temperature
5) Great room temperature
6) Pump water pressure

One PLC output will be used to drive a horn. A second output will be used to drive a blower fan under the control of the Visual Basic program. The fan will turn on when the temperature detected in the great room indicates that a wood-burning stove is on, or if there is significant heat due to solar radiation through the windows. When the fan is on, the hot air collected from the great room will be routed to cooler section of the house. This process redistributes the excess heat, provides a more uniform heat distribution, and saves energy.

For the home monitoring task, an Allen-Bradley MicroLogix PLC model number 1761-L20BWB-5A will be used. This model has 12 discrete 24V DC inputs, eight relay outputs, four analog inputs, and one analog output. The four analog inputs are composed of two voltage inputs (with a range of 0 to 10 volts) and two current inputs (with a range of 4 to 20 milliamps). Although a 120V AC model is available, this particular model is DC-powered. Therefore, an external 24V DC power supply is required. The physical dimensions of this unit are 7.87 inches by 3.15 inches by 1.57 inches. A photograph of this PLC, provided courtesy of Allen-Bradley, is presented in Figure 6.01.

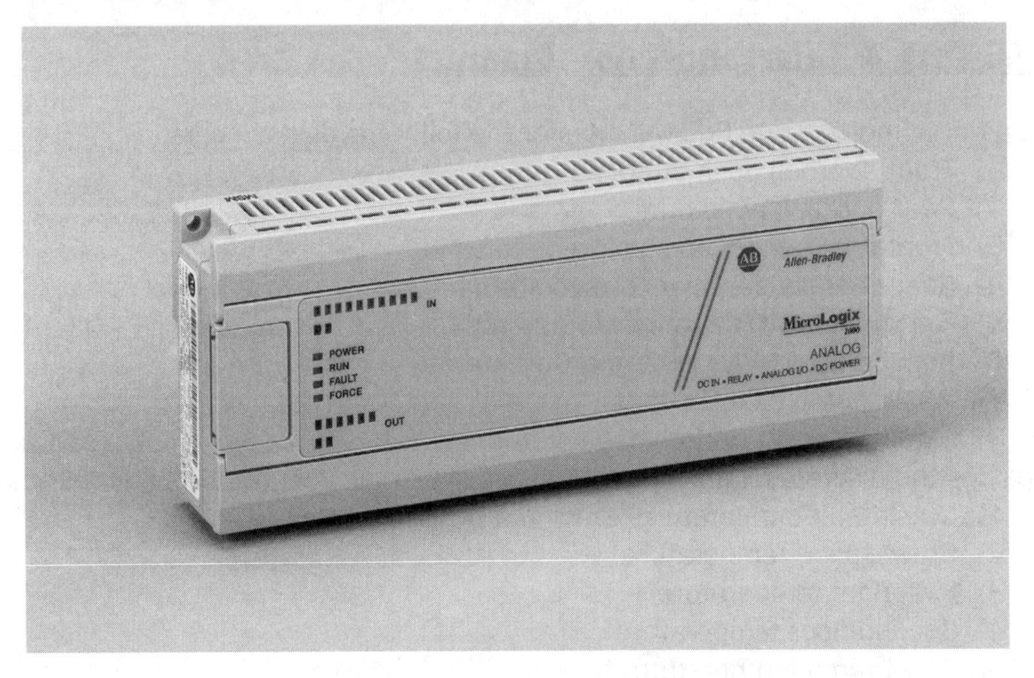

Figure 6.01

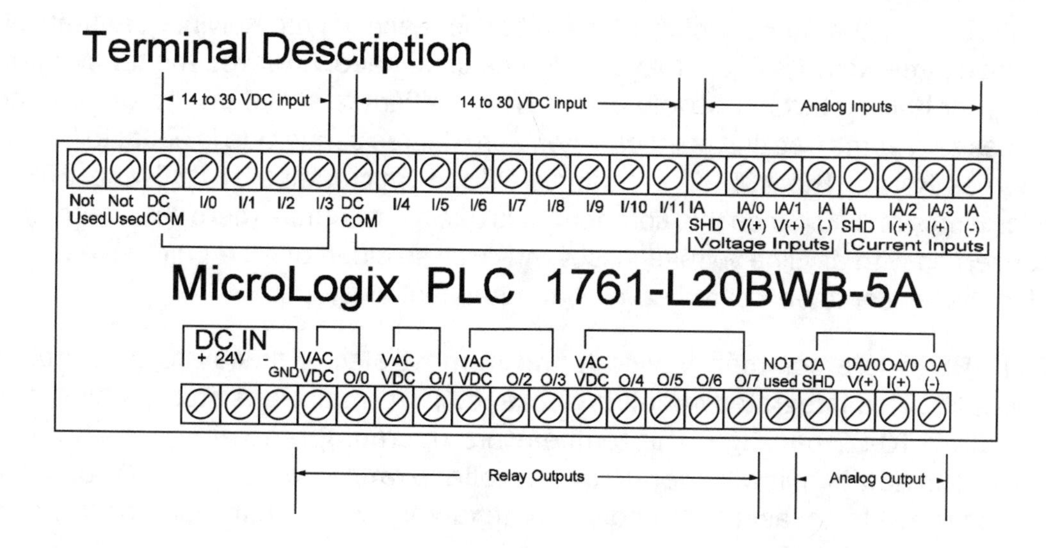

Figure 6.02

Figure 6.02 provides the terminal layout for this MicroLogix PLC. There are two electrically isolated groups of discrete inputs. If desired, you can interface to different systems and maintain electrical isolation. An analog input group is available with two voltage inputs and two current inputs. Four electrically isolated groups of relay outputs are available. The output grouping permits interfacing to different voltages. An analog output group provides a single analog output. The analog output can be voltage or current.

The project requirement of eight discrete inputs is easily satisfied by this PLC, which has 12 inputs. Four analog inputs are available, but six analog inputs are required. This presents a small problem. One way to eliminate this problem is to use a technique called multiplexing. Multiplexing allows multiple devices to share one resource. In this situation, the shared resource is an analog input channel. Five temperature readings need to be collected. Since temperature is normally a rather slow-moving event, it's the perfect candidate to be multiplexed; however, before you develop a multiplexing scheme, you first need to choose a temperature sensor for the task.

Analog Devices, Inc. makes a temperature-controlled, current source-integrated circuit called an AD590. This two-terminal device is based on the Kelvin temperature scale. One microampere of current is emitted for each degree Kelvin. Quantitatively, a unit degree Kelvin is equal to a unit degree centigrade; however, the temperature scale for Kelvin starts at absolute zero. The centigrade scale, on the other hand, places zero at the freezing point of water. Zero degree centigrade is equal to 273.15 degrees Kelvin. The AD590 will source 273.15 microamps at 0°C and 274.15 microamps at 1°C. Since this component is a current source, there's no need to be concerned with any line voltage losses. Any well-shielded twisted pair cable will enable the sensor to be located several hundred feet away.

The AD590 is a common device found in most electronic catalogs and costs about $5. This integrated circuit is packaged in a small transistor housing, specifically a TO-52 package. The temperature operating range is from –55°C to 150°C. In addition, it will operate over a voltage range from +4 to +30 volts. The device is available laser-trimmed from the factory with a calibration accuracy of plus or minus 0.5°C.

The AD590 does not source enough current to be used with a 4- to 20-milliamp input. A resistor will be used to convert the current output of the AD590 into voltage, in order to be read by a voltage input. If we limit the temperature to 122°F or 50°C, the theoretical current emitted from the AD590 is 323.15 microamps. The value for this resistance is determined by Ohm's law. The maximum input voltage of 10 divided by 323.15 microamps produces a resistance of 30,945 ohms. A 30.9-kohm resistor will be used for the conversion.

A quick check of the input impedance for the analog-voltage input reveals a small problem. The input impedance is specified at 210 kohms. This impedance is small enough to significantly alter the value of the temperature resistor. A 36.5-kohm resistor in parallel with a 210-kohm resistor produces an effective resistance of 31,095 ohms. This value will be used to convert the voltage reading to a temperature value.

Since the outside temperature is the most dynamic, the outside temperature sensor will be dedicated to the first analog voltage input. The remaining temperature sensors will share the second analog voltage input. Micrologix outputs 0/4, 0/5, 0/6, and 0/7 will be used to implement the multiplexing scheme. Figure 6.03 illustrates the scheme.

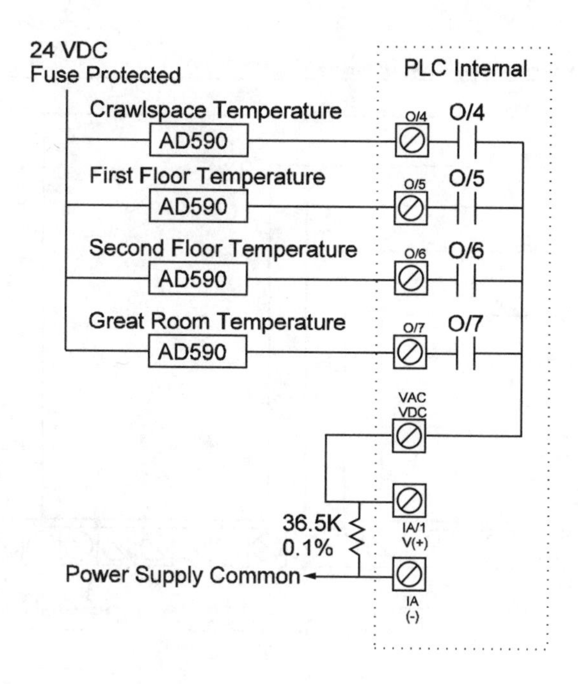

Figure 6.03

Standard door push buttons will be used for the doorbell. Magnetic reed proximity switches—with a companion magnetic—will be used to monitor the open/close status of the front door, rear door, and mailbox door. The first-floor and second-floor HVAC units will be monitored with the optical-isolator circuit described in Chapter 4. A relay with a 240V AC coil is used to monitor the water pump on/off status.

Water pressure will be monitored with a 4- to 20-milliamp 100-psi pressure transmitter. This pressure transmitter is housed with stainless steel. Factory calibration provides an accuracy of plus or minus 0.5%.

In addition, the pressure transducer is fully temperature-compensated. A voltage-operating range between 12 V DC and 36 V DC is required for proper operation. An interconnection diagram between the required components and the Allen-Bradley MicroLogix PLC is illustrated in Figure 6.04.

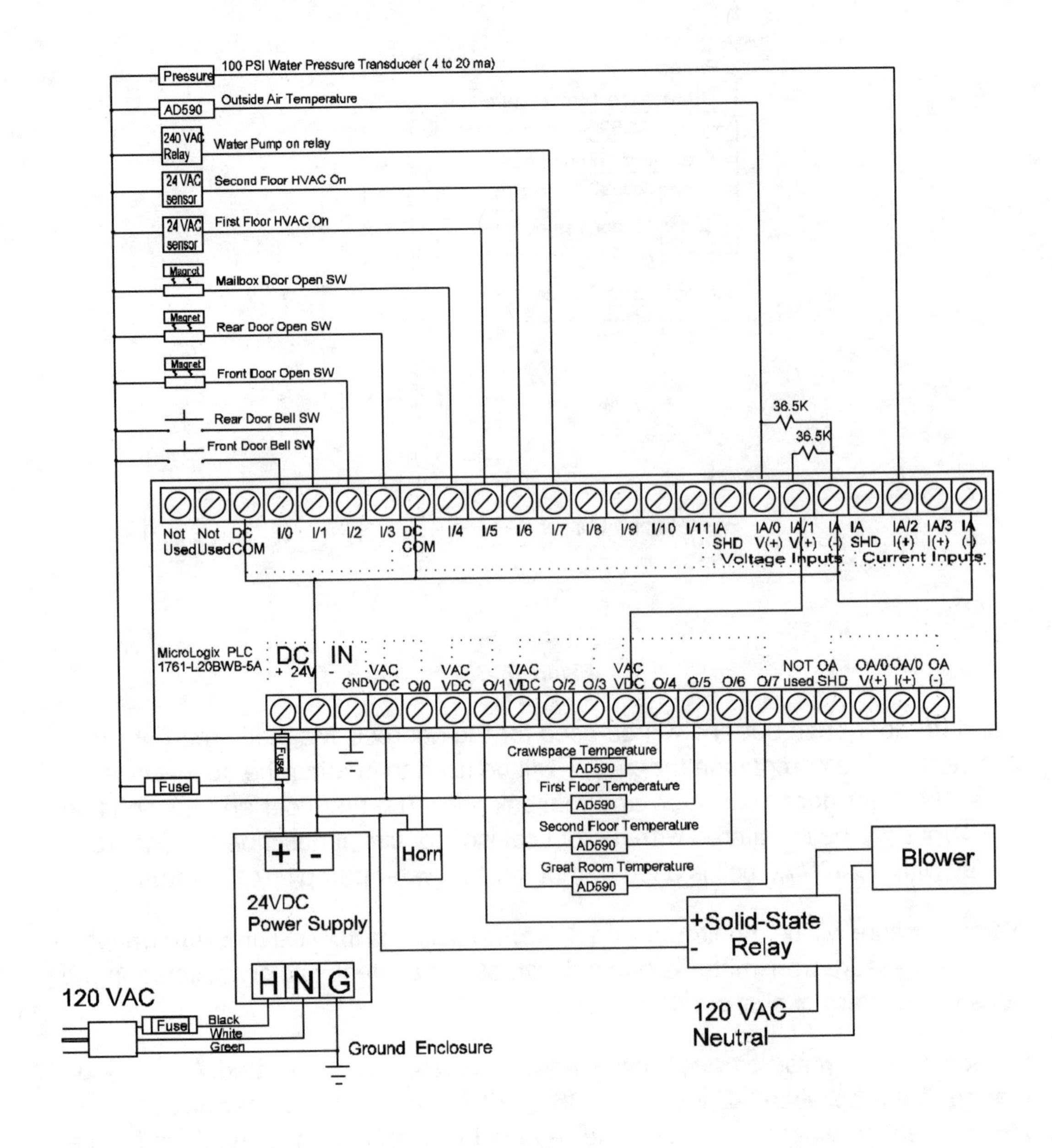

Pressure
100 PSI Water Pressure Transducer (4 to 20 ma)
AD590
Outside Air Temperature
240 VAC Relay
Water Pump on relay
24 VAC sensor
Second Floor HVAC On
24 VAC sensor
First Floor HVAC On
Mailbox Door Open SW
Rear Door Open SW
Front Door Open SW
Rear Door Bell SW
Front Door Bell SW
36.5K
36.5K
Not Used
Not Used
DC COM
I/0
I/1
I/2
I/3
DC COM
I/4
I/5
I/6
I/7
I/8
I/9
I/10
I/11
IA SHD
IA/0 V(+)
IA/1 V(+)
IA (-)
IA SHD
IA/2 I(+)
IA/3 I(+)
IA (-)
Voltage Inputs
Current Inputs
MicroLogix PLC
1761-L20BWB-5A
DC IN
+ 24V -
GND
VAC VDC
O/0
VAC VDC
O/1
VAC VDC
O/2
O/3
VAC VDC
O/4
O/5
O/6
O/7
NOT used
OA SHD
OA/0 V(+)
OA/0 I(+)
OA (-)
Fuse
Fuse
Crawlspace Temperature
AD590
First Floor Temperature
AD590
Second Floor Temperature
AD590
Great Room Temperature
AD590
Horn
24VDC
Power Supply
H N G
120 VAC
Black
White
Green
Ground Enclosure
Blower
+Solid-State
- Relay
120 VAC
Neutral

Figure 6.04

Ladder Logic Software

The required ladder logic for this project is not very complicated. For the most part, the PLC is acting as an input/output interface. The ladder logic tasks are as follows.

1) The PLC's timers will be utilized to perform filtering of the input signal. This filtering action ensures signal integrity.
2) The temperature multiplexing ladder logic will be developed.
3) The PLC will keep track of the daily run time and the daily number of cycles for the HVAC systems and the water pump. A bit will be sent at midnight from the corresponding Visual Basic program to clear all the daily data registers for the new day.
4) Since the computer is initiating all the communications, a bit called a "ComBit" will alternate states with each read/write command from the computer. The dynamic nature of this bit will indicate to the PLC that a good communication link exists with the PC computer. The horn will then be used for various annunciation functions when communication with the computer stops. As part of this home automation project, specific wave files will be launched for the various audio-annunciation tasks. These tasks include the doorbell, door open beep, and new mail. The horn is a backup that is PLC-driven.

Figure 6.05 illustrates the software required for filtering the discrete inputs. The purpose of the filtering is to eliminate the multiple transitions associated with switch bounce and to produce a signal-pulse width that is of a long enough duration to be picked up by the computer's communications routines. When the front door push button is pressed continuously for 0.1 second, timer T4:0 times up to the preset value, timing stops, and bit T4:0/DN is asserted. T4:0/DN latches bit N7:0/0 and resets timer T4:1. Bit N7:0/0 is collected by the computer's serial acquisition. When the front door push button is released, timer T4:1 begins timing and stops when the preset time of 0.5 is reached. Bit N7:0/0 is cleared (unlatched) by the assertion of T4:1/DN. Timer T4:1 produces a hold delay of 0.5 second.

This code is duplicated seven times for each of the discrete input signals. The timer addresses and computer acquisition-bit addresses follow a normal numerical

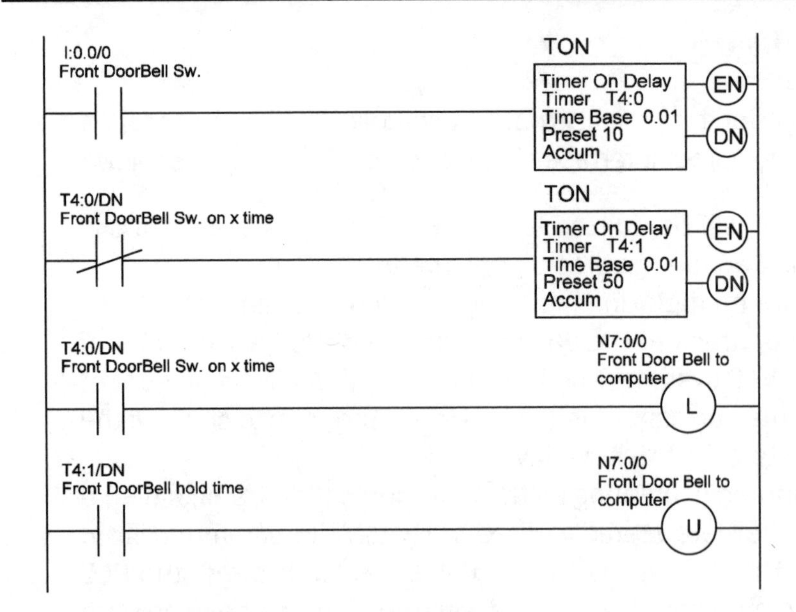

Figure 6.05

Discrete Input Description	*PLC input Address*	*Delay Timer Address*	*Hold Timer Address*	*Computer Acquisition Bit Address*
Front DoorBell Switch	I:0.0/0	T4:0	T4:1	N7:0/0
Rear DoorBell Switch	I:0.0/1	T4:2	T4:3	N7:0/1
Front Door Open Switch	I:0.0/2	T4:4	T4:5	N7:0/2
Rear Door Open Switch	I:0.0/3	T4:6	T4:7	N7:0/3
MailBox Door Open Switch	I:0.0/4	T4:8	T4:9	N7:0/4
First Floor HVAC On	I:0.0/5	T4:10	T4:11	N7:0/5
Second Floor HVAC On	I:0.0/6	T4:12	T4:13	N7:0/6
Water Pump On	I:0.0/7	T4:14	T4:15	N7:0/7

Table 6.01

progression for each input. Table 6.01 illustrates the required addressing for each input.

The next section of ladder logic performs the temperature-multiplexing scheme required to collect all the specified temperatures. The required software is illustrated in Figures 6.06 and 6.07. With every minute of time, a different temperature probe is energized. The probe selection process is determined by register

N7:104, which increments with each positive transition of timer DN-bit of T4:16. Once T4:16/DN goes true, the timer T4:16 clears along with T4:16/DN and a new timing cycle starts. Register N7:104 resets back to a value of zero as soon as the incremental count reaches four. Therefore, the counting pattern of register N7:104 is 0, 1, 2, 3, 0, 1, and so on. The ladder logic software presented in Figure 6.06 performs this task.

One potentially weak link in this application is that a relay contact problem with respect to conductivity may develop over time, because the current demand through the relay contact is such a small value. This current is less than 300 microamps. Allen-Bradley specifies a minimum relay-contact current of 10 milliamps. A minimum amount of current helps to keep the contacts polished and conducting properly. If a problem does arise, relays with contacts designed to drive small currents loads, such as this temperature sensor, can be driven by these contacts. The coil load of these relays will meet the 10-milliamp minimum contact-current requirement.

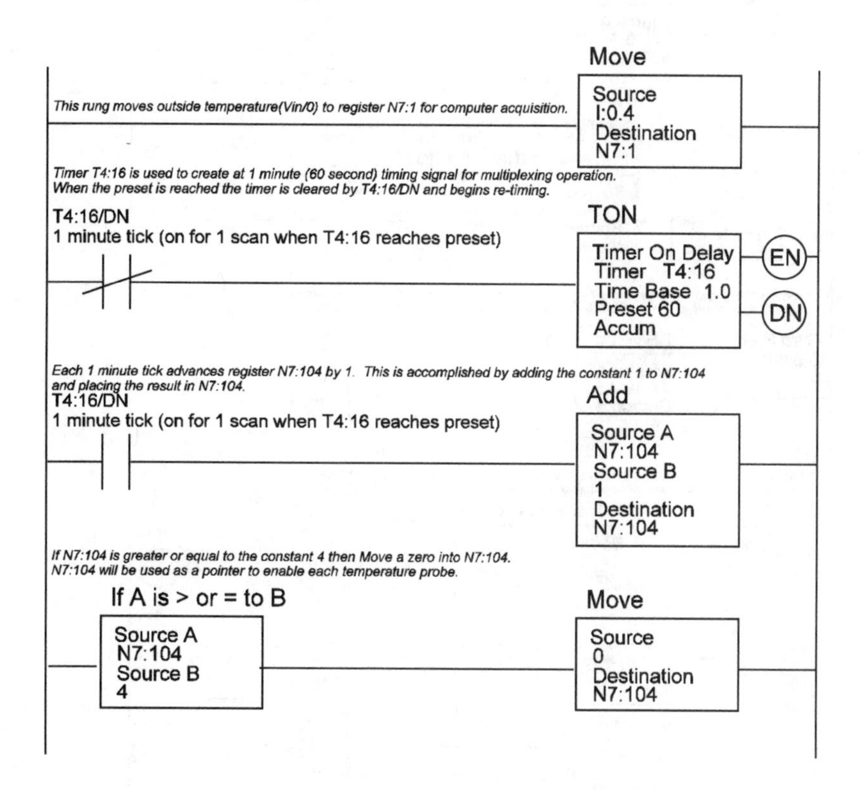

Figure 6.06

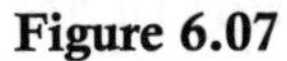
Figure 6.07

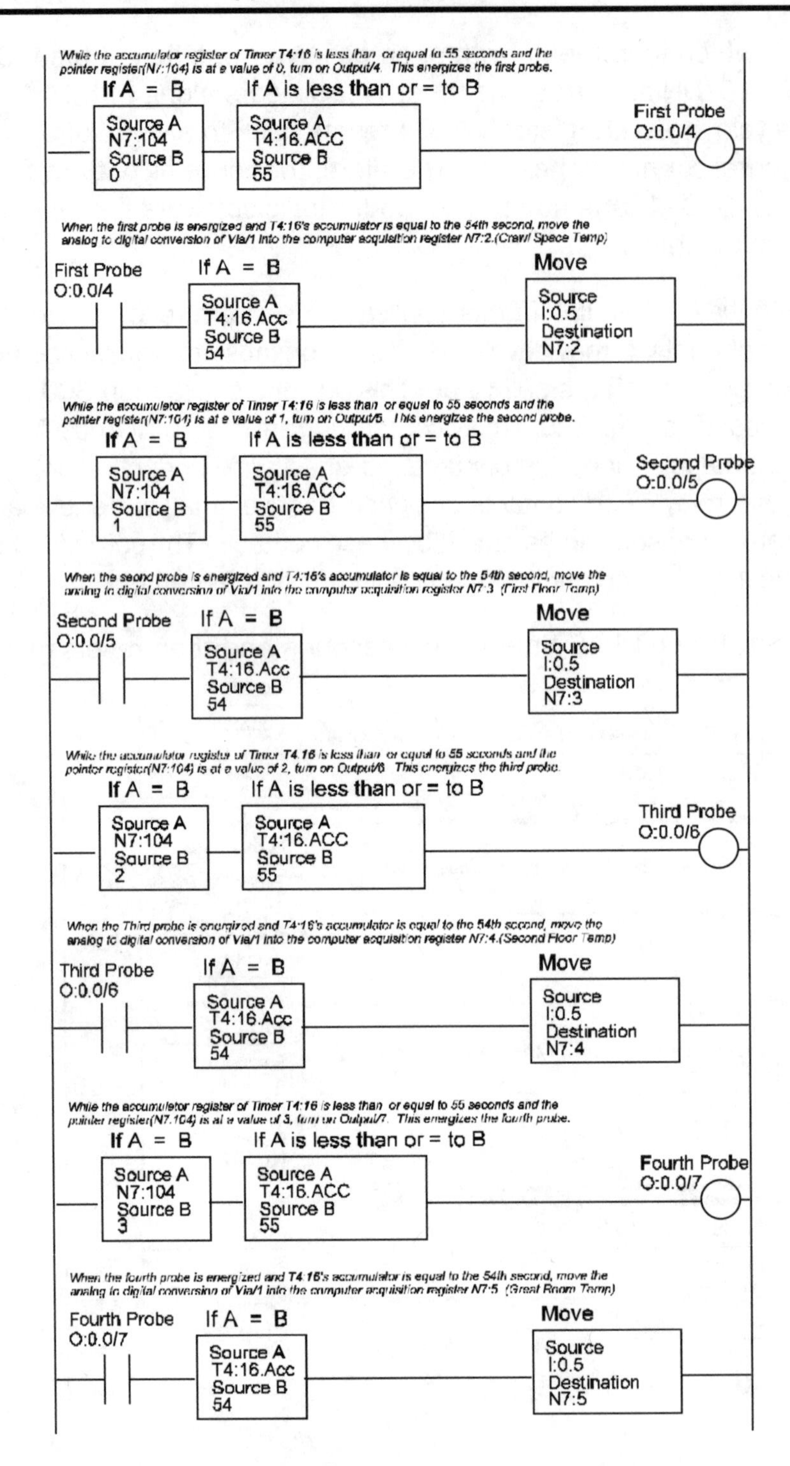
While the accumulator register of Timer T4:16 is less than or equal to 55 seconds and the pointer register(N7:104) is at a value of 0, turn on Output/4. This energizes the first probe.
If A = B
If A is less than or = to B
Source A
N7:104
Source B
0
Source A
T4:16.ACC
Source B
55
First Probe
O:0.0/4
When the first probe is energized and T4:16's accumulator is equal to the 54th second, move the analog to digital conversion of Via/1 into the computer acquisition register N7:2.(Crawl Space Temp)
First Probe
O:0.0/4
If A = B
Source A
T4:16.Acc
Source B
54
Move
Source
I:0.5
Destination
N7:2
While the accumulator register of Timer T4:16 is less than or equal to 55 seconds and the pointer register(N7:104) is at a value of 1, turn on Output/5. This energizes the second probe.
If A = B
If A is less than or = to B
Source A
N7:104
Source B
1
Source A
T4:16.ACC
Source B
55
Second Probe
O:0.0/5
When the seond probe is energized and T4:16's accumulator is equal to the 54th second, move the analog to digital conversion of Via/1 into the computer acquisition register N7:3 (First Floor Temp)
Second Probe
O:0.0/5
If A = B
Source A
T4:16.Acc
Source B
54
Move
Source
I:0.5
Destination
N7:3
While the accumulator register of Timer T4:16 is less than or equal to 55 seconds and the pointer register(N7:104) is at a value of 2, turn on Output/6. This energizes the third probe.
If A = B
If A is less than or = to B
Source A
N7:104
Source B
2
Source A
T4:16.ACC
Source B
55
Third Probe
O:0.0/6
When the Third probe is energized and T4:16's accumulator is equal to the 54th second, move the analog to digital conversion of Via/1 into the computer acquisition register N7:4.(Second Floor Temp)
Third Probe
O:0.0/6
If A = B
Source A
T4:16.Acc
Source B
54
Move
Source
I:0.5
Destination
N7:4
While the accumulator register of Timer T4:16 is less than or equal to 55 seconds and the pointer register(N7:104) is at a value of 3, turn on Output/7. This energizes the fourth probe.
If A = B
If A is less than or = to B
Source A
N7:104
Source B
3
Source A
T4:16.ACC
Source B
55
Fourth Probe
O:0.0/7
When the fourth probe is energized and T4:16's accumulator is equal to the 54th second, move the analog to digital conversion of Via/1 into the computer acquisition register N7:5 (Great Room Temp)
Fourth Probe
O:0.0/7
If A = B
Source A
T4:16.Acc
Source B
54
Move
Source
I:0.5
Destination
N7:5

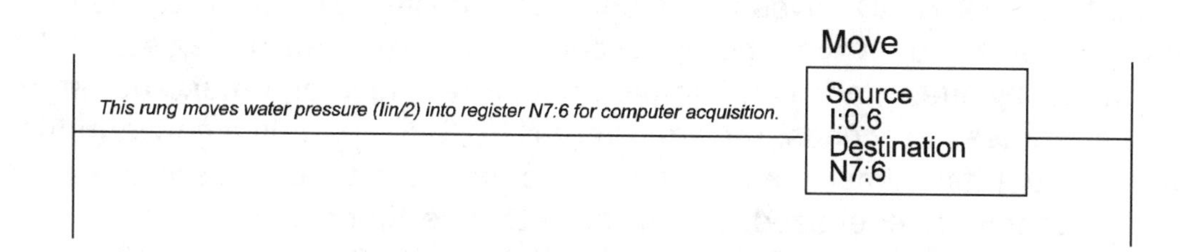
Move
This rung moves water pressure (Iin/2) into register N7:6 for computer acquisition.
Source
I:0.6
Destination
N7:6

Figure 6.08

The active temperature probe's current output creates a voltage across the 30.9-kohm resistor at the second analog-voltage input V(+)Ia/1. This voltage is proportional to temperature. Fifty-four seconds is allocated to allow the temperature-probe signal to stabilize. During the 54th second, the result of the analog-to-digital conversion is moved in the appropriate computer-acquisition register. When the accumulator register of timer T4:16 exceeds 55 seconds, the active probe is de-energized. Timer T4:16 times to 60 seconds and then resets to zero. The probe pointer N7:104 advances, and a new cycle is set in motion.

Figure 6.07 provides the ladder logic software to acquire the multiplexed-temperature data. Based on the value of the pointer register N7:104 and the multiplex timer's accumulator register (T4:16.Acc), the appropriate probe is energized for 54 seconds and then an analog-to-digital snapshot is taken. The acquired temperature value is stored in the appropriate register.

The water-pressure transducer is wired to I(+)inA/2. This analog input accepts a 0- to 20-milliamp input. Figure 6.08 shows the rung of ladder logic that moves the analog-to-digital result to computer-acquisition register N7:6. The next section of ladder logic implements the code required for the daily number of cycles and daily-accumulated on-time tasks. Both HVAC systems and the water pump are monitored for daily cycles and run time. These functions could have been implemented with Visual Basic; however, the PLC coding of these functions requires less code and is more effective.

The ladder logic illustrated in Figure 6.09 keeps track of the daily number of cycles and daily run time for the first-floor HVAC unit. The conditioned on-bit for the first-floor HVAC drives a one-shot relay hosted by bit B3/1. The one-shot function ensures that the cycle counter constructed by register N7:7 increments by one. Timer T4:17 implements a retentive on-timer. The retentive timer times in seconds, provided that N7:0/5 is on. Unlike conventional timers, a retentive timer maintains the accumulated time value if the rung should become de-energized. When the retentive timer reaches the preset value of 60, timer bit T4:17/DN is asserted. T4:17/DN advances the first-floor HVAC minute-totalizer register N7:8. A transition of T4:17/DN, or the computer-generated reset, B3/0 causes the retentive timer T4:17 to clear. Both register N7:7 and register N7:8 are in the computer-acquisition area.

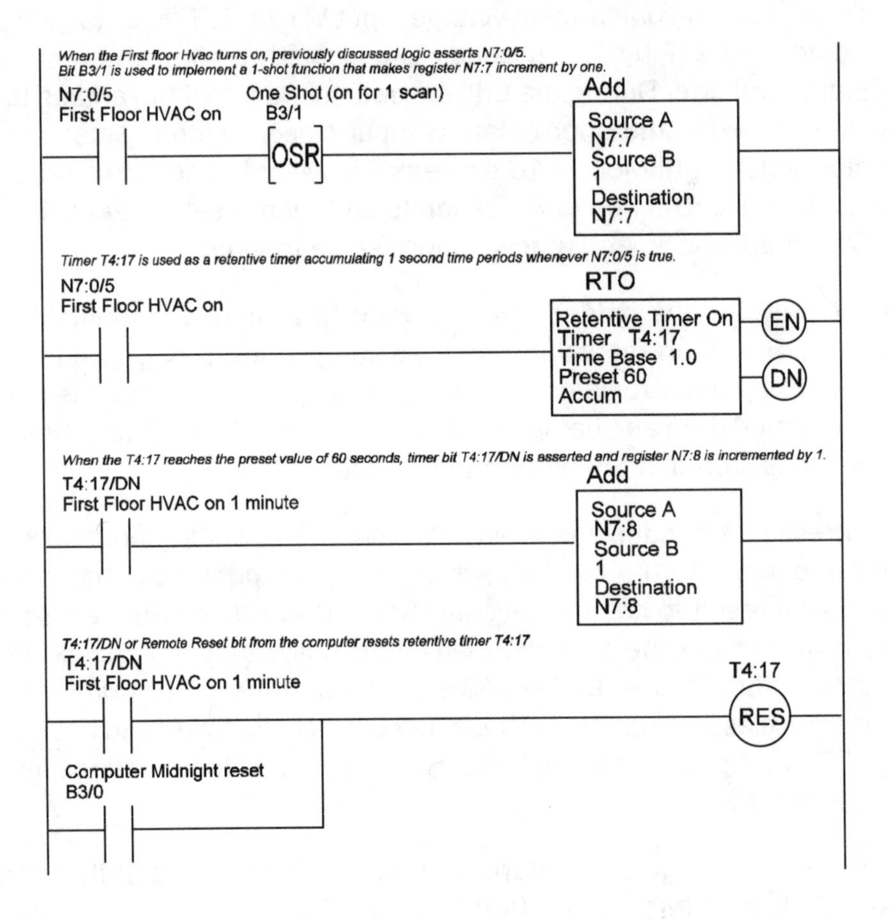

Figure 6.09

The daily cycle counter and daily run-time ladder logic code is duplicated for both the conditioned second-floor HVAC signal (N7:0/6) and the water-pump signal (N7:0/7). Table 6.02 shows the allocated registers, timers, and bits required for the cycle count and run-time tasks.

Unit to Monitor for number of cycles and run time	*PLC conditioned output*	*1 shot bit address*	*Cycle Counter register*	*Retentive timer address*	*Accumulated run time register*
First Floor HVAC	N7:0/5	B3/1	N7:7	T4:17	N7:8
Second Floor HVAC	N7:0/6	B3/2	N7:9	T4:18	N7:10
Water Pump	N7:0/7	B3/3	N7:11	T4:19	N7:12

Table 6.02

The ladder logic provided in Figure 6.10 constructs the communication-fail horn-driver. Figure 6.11 shows the software required for the computer-generated reset of daily registers. At midnight, the PC sets N7:50 bit 4 high. The PLC receives this

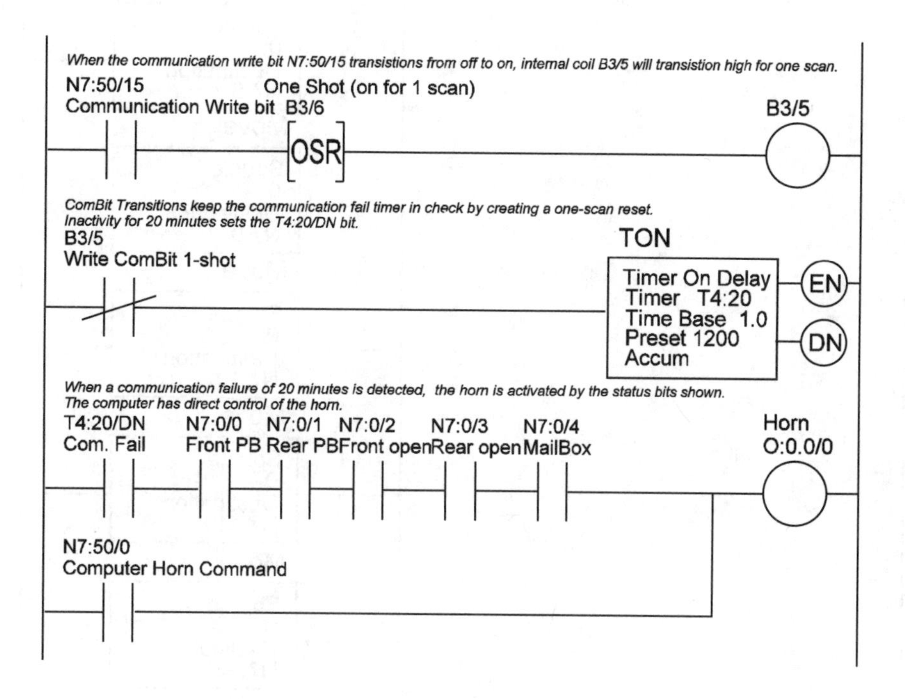

Figure 6.10

bit and drives a one-shot coil, which clears all of the cycle-counter and run-time registers. The one-shot coil output is high for only 1 PLC scan time. Figure 6.12 shows the required ladder logic to implement the computer-output control. With the ladder logic software complete and in place, it's time to move on to the Visual Basic parts of the home-monitor project. Table 6.03 summarizes the location and description of the required data acquisition.

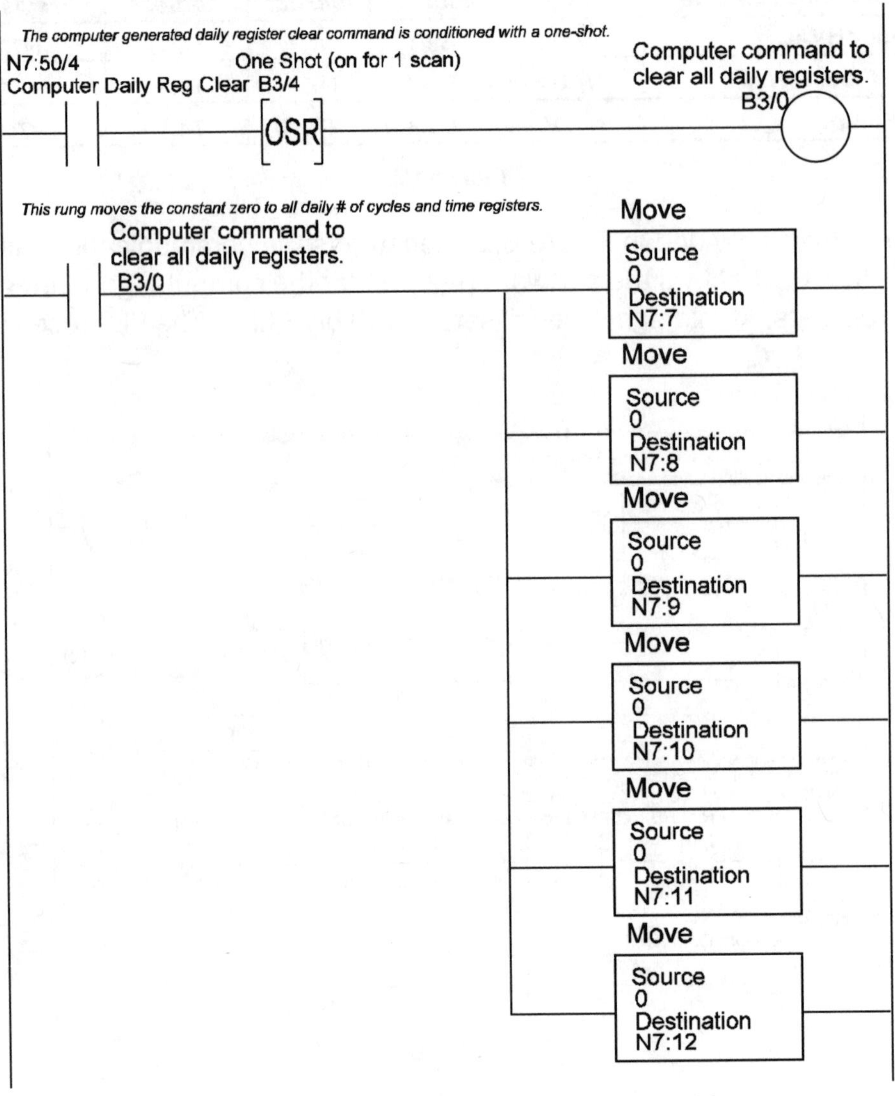

Figure 6.11

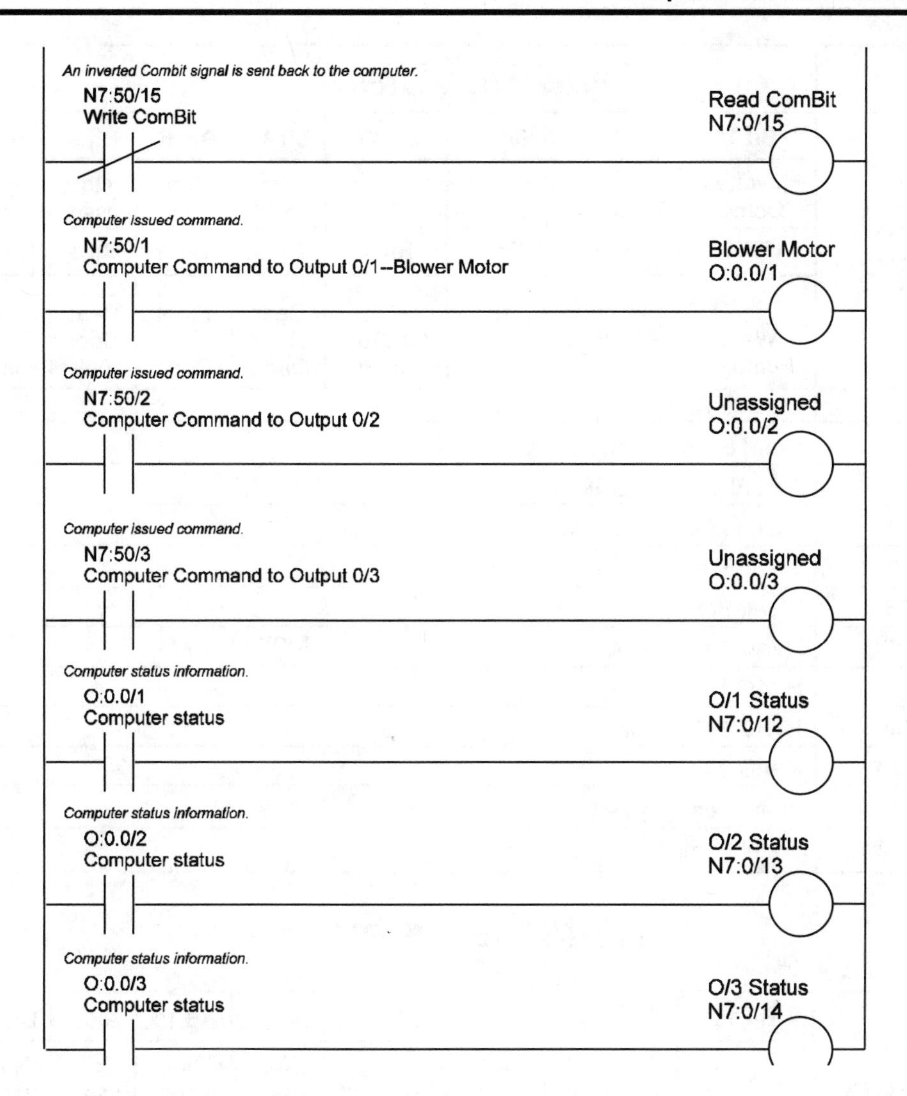
An inverted Combit signal is sent back to the computer.
N7:50/15
Write ComBit
Read ComBit
N7:0/15
Computer issued command.
N7:50/1
Computer Command to Output 0/1--Blower Motor
Blower Motor
O:0.0/1
Computer issued command.
N7:50/2
Computer Command to Output 0/2
Unassigned
O:0.0/2
Computer issued command.
N7:50/3
Computer Command to Output 0/3
Unassigned
O:0.0/3
Computer status information.
O:0.0/1
Computer status
O/1 Status
N7:0/12
Computer status information.
O:0.0/2
Computer status
O/2 Status
N7:0/13
Computer status information.
O:0.0/3
Computer status
O/3 Status
N7:0/14

Figure 6.12

	Read Registers							
N7:0 BITS	**Bit 15**	**Bit 14**	**Bit 13**	**Bit 12**	**Bit 11**	**Bit 10**	**Bit 9**	**Bit 8**
	Inverted ComBit	Status O/3	Status O/2	Status O/1	Not used	Not used	Not used	Not used
	Bit 7	**Bit 6**	**Bit 5**	**Bit 4**	**Bit 3**	**Bit 2**	**Bit 1**	**Bit 0**
	Water Pump On	2F HVAC On	1F HVAC On	MailBox Open	Rear Door Open	Front Door Open	Rear Door Bell	Front Door bell
N7:1	Outside Temperature							
N7:2	Crawl Space Temperature							
N7:3	First Floor Temperature							
N7:4	Second Floor Temperature							
N7:5	Great Room Temperature							
N7:6	Water Pressure							
N7:7	HVAC 1 Cycle Count							
N7:8	HVAC 1 Run Time							
N7:9	HVAC 2 Cycle Count							
N7:10	HVAC 2 Run Time							
N7:11	Water Pump Cycle Count							
N7:12	Water Pump Run Time							
	Write Registers							
N7:50 BITS	**Bit 15**	**Bit 14**	**Bit 13**	**Bit 12**	**Bit 11**	**Bit 10**	**Bit 9**	**Bit 8**
	ComBit	Not used	Not used	Not used	Not used	Not used	Not used	Not used
	Bit 7	**Bit 6**	**Bit 5**	**Bit 4**	**Bit 3**	**Bit 2**	**Bit 1**	**Bit 0**
	Not used	Not used	Not used	Daily Register Clear	O/3 CMD	O/2 CMD	O/1 CMD	Horn Cmd

Table 6.03

Chapter 7
HOME-MONITOR ACQUISITION MODULE

INTRODUCTION

Now that the PLC details of the required data acquisition have been taken care of in Chapter 6, it's time to move on to the actual data acquisition and animation aspects of the home-monitor project. The Visual Basic programming tasks for this project can be broken down in two parts: the data-acquisition software and the animation software.

Task Description

The data-acquisition task is the focus of this chapter; Chapter 8 will cover the animation software. The acquisition software developed in Chapter 5 is very similar to this acquisition-software requirement. It's so similar, in fact, that I considered simply highlighting where the changes occur. But I decided instead that the best method is to provide the clearest details of the acquisition task. If you didn't understand the acquisition on the first pass in Chapter 5, then hopefully the second pass in this chapter will drive the concept home.

Similar to the software in Chapter 5, the PC is initiating the polling requests to the PLC. The requirements of this acquisition task, however, call for an alternating read/write request. In addition, a designated bit called a ComBit is used to provide the PLC with communication information. This bit changes state with each successful read/write transaction. A communication failure is detected by the ComBit remaining in one state. If a communication failure is detected by the PLC, the PLC will use a horn output for any of the required audio tasks. Refer back to Table 6.03 in the previous chapter for a summary of the PLC data that needs to be acquired.

The acquisition software will acquire the register information and break out the data into Visual Basic variables. The analog data will be mathematically converted into temperature units and pressure units. The temperature probes may require calibration. Provisions will be made for temperature-probe calibration. The pressure transducer has been laser-trim calibrated at the factory, so calibration will not be required. Lastly, the normalized data will be compiled and stored in a file on the hard drive. If you have a local area network, the other computers on the network can access this file and provide local animation. This project will limit this feature to read-only access. Now it's time to start coding. The following code was created by modifying the code developed in Chapter 5. The complete code is provided on the companion CD-ROM. You will begin with variable declaration, which is located in a module called module 1. If you are starting with a new project, then you will need to add a module. If you are modifying the code that you created in Chapter 5, then now is the time to do a file-save-as for this new project. You may not want to lose the code you created in Chapter 5.

The Code

Place the following code in the general-declaration section of Module 1.

```
Option Explicit

'COMMUNICATION VARIABLES AND CONSTANTS

Public Const STX As Byte = 2
Public Const ETX As Byte = 3
Public Const DLE As Byte = 16
Public Const ACK As Byte = 6
Public Const ENQ As Byte = 5
Public Const NAK As Byte = 21
Public Const STS As Byte = 0
```

```
Public bDST As Byte 'PLC Address
Public bSRC As Byte 'PC address always zero
Public bCMD As Byte 'command byte
Public bLCRC As Byte 'calculated low byte CRC
Public bHCRC As Byte 'calculated high byte CRC
Public bRxLCRC As Byte 'Received low byte of CRC
Public bRxHCRC As Byte 'Received high byte of CRC


Public bDATA As Byte   ' first byte of write command
Public bDATA2 As Byte   ' second byte of write command
Public bLastLastRX As Byte 'last last receive byte
Public bLastRX As Byte  ' last receive byte
Public bNewRx As Byte  'newest receive byte
Public bRXpointer As Byte 'used for receive pointer
Public bETXposition As Byte 'location of ETX
Public STXflag As Boolean 'indicates an STX was detected
Public ETXflag As Boolean 'indicates an ETX was detected
Public ReadModeFlag As Boolean 'used to differentiate between a read
and a write request
Public bComStep As Byte 'used to show four steps of communication
Public bReadNumberOfRegisters As Byte ' the number of registers to
acquire

Public bPlcData(200) As Byte 'storage area for PLC register bytes
Public FrontDoorBellflag As Boolean 'true indicates Front door bell pushed
Public RearDoorBellflag As Boolean 'true indicates rear door bell pushed
Public FrontDoorOpenflag As Boolean 'true indicates the front door is open
Public RearDoorOpenflag As Boolean 'true indicates the rear door is open
Public MailBoxflag As Boolean 'true indicates the mailbox is open
Public FirstFloorHvacOnflag As Boolean 'true indicates the 1f hvac is on
Public SecondFloorHvacOnflag As Boolean 'true indicates the 2f hvac is on
Public WaterPumpOnflag As Boolean 'true indicates the water pump is on
Public ComBitflag As Boolean 'used to indicate good communication PC to PLC
Public StartUpflag As Boolean 'initialization flag
Public Status01flag As Boolean 'PLC output 0/1
```

```
Public Status02flag As Boolean 'PLC output O/2
Public Status03flag As Boolean 'PLC output O/3

Public lRawOutsideTemp As Long 'Analog to Digital value
Public lRawCrawlSpaceTemp As Long 'Analog to Digital value
Public lRawFirstFloorTemp As Long 'Analog to Digital value
Public lRawSecondFloorTemp As Long 'Analog to Digital value
Public lRawGreatRoomTemp As Long 'Analog to Digital value
Public lRawWaterPressure As Long 'Analog to Digital value
Public lHvac1DailyCycle As Long 'number of times called during the day
Public lHvac1DailyOnTime As Long 'total on time in minutes
Public lHvac2DailyCycle As Long 'number of times called during the day
Public lHvac2DailyOnTime As Long 'total on time in minutes
Public lWaterPumpDailyCycle As Long 'number of times called during the day
Public lWaterPumpDailyOnTime As Long 'total on time in minutes
Public vOutsideTemp  'Temperature in degrees F
Public vCrawlSpaceTemp 'Temperature in degrees F
Public vFirstFloorTemp 'Temperature in degrees F
Public vSecondFloorTemp 'Temperature in degrees F
Public vGreatRoomTemp 'Temperature in degrees F
Public vWaterPressure ' Pressure in PSI

Public vTxtData 'used to store transmitted data
Public vRxData 'used to store received data

'Write Bits to the PLC
Public HornCmdFlag As Boolean 'drives the horn O/0
Public Out1CmdFlag As Boolean 'drives O/1
Public Out2CmdFlag As Boolean 'drives O/2
Public Out3CmdFlag As Boolean 'drives O/3
Public ClearDailyRegCmdFlag As Boolean 'clears all daily data

Public Const lCRCconstant As Long = 40961 ' HEX  A001
Public lTransaction As Long 'used to provide a transaction number

Public bTxData(100) As Byte 'byte array used for transmit data
Public bRxData(100) As Byte 'byte array used for receive data
```

Add the following Sub procedure to this module. This Sub procedure is used for all CRC calculation.

```
Public Sub COMPUTECRC()

Dim CrcRegister As Long 'this is just a scratchpad variable specific to this sub
 Dim T As Byte
 Let T = bTxData(0) + 1 'VB6 always allocates array(0) because the default lower
                    'bound is zero unless an "Option Base 1" statement is
                    'placed in the declaration section of a module. Let's use
                    'this byte to point to the end of the data in the array.

Let bTxData(T) = ETX 'LOAD IN AN ETX AS REQUIRED FOR CRC CALCULATION
AFTER THE LAST DATA BYTE IN THE ARRAY.
 Dim D As Byte
 Dim ShiftFLAG As Boolean

  Let CrcRegister = 0

 For T = 1 To (bTxData(0) + 1) 'FOR ALL DATA INCLUDING THE ADDED ETX
   Let CrcRegister = Int(CrcRegister Xor bTxData(T))
       For D = 1 To 8
     'If the least significant bit is high this is equivalent to shift out so perform
       'an Exclusive Or of the CrcRegister with the CRCconstant
     If (CrcRegister And 1)=1 Then Let ShiftFLAG=True Else Let ShiftFLAG=False
      Let CrcRegister = Int(CrcRegister / 2) 'The Int keyword removes the
fractional part of the number and returns the resulting integer value.
This prevents rounding which distorts the result.
     If ShiftFLAG = True Then Let CrcRegister = CrcRegister Xor lCRCconstant
     Next D
 Next T

Let bLCRC = CrcRegister And 255 'low byte of CRC
'find the high byte of the CRC
Let bHCRC = Int(CrcRegister / 256) ' Use the Int keyword to prevent rounding
End Sub
```

This completes the software requirements of the Module. Now move on to Form1. Form1 will be used to coordinate the communications between the PC and the PLC. The received data will be broken down into Visual Basic variables. It's not necessary for this form to be visible, so the default load is Form1.visible= false. The layout of Form1 is illustrated in Figure 7.01.

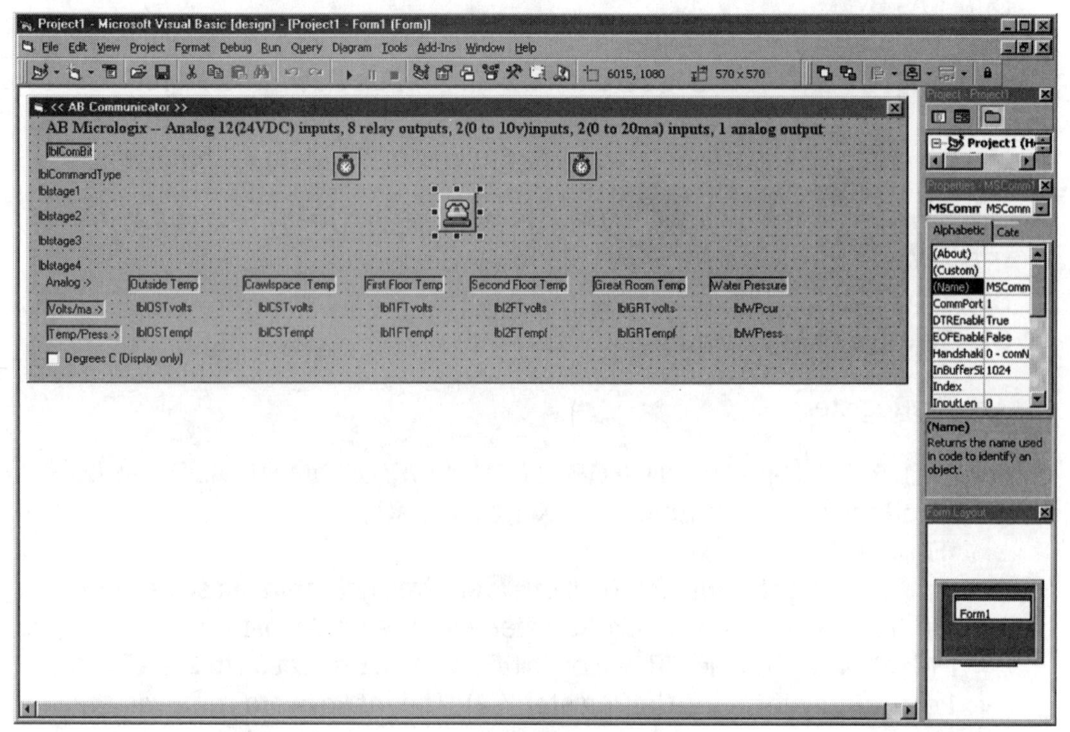

Figure 7.01

The form is configured with a fixed single border and a caption of "<< AB Communicator >>". The fixed single border keeps the form size fixed and eliminates the MinButton and the MaxButton. There is no need to adjust the size of this form.

Two timers are provided. The first timer is named "READtimer" and is configured with an interval of 75, representing 75 milliseconds. The second timer is named "ComTimeout" and is configured with an interval of 250 or 250 milliseconds.

An MsComm object is also placed on the form to handle the serial communications. MsComm is not a standard toolbox tool and will need to be added to the toolbox. This is accomplished by right-clicking on the toolbox and selecting components with a left click. Scroll down the controls list box and place a check in the checkbox of "Microsoft Comm Control 6.0".

The remaining controls are all labels. A close up view of the labels contained on Form1 is presented Figure 7.02.

All of the labels under software control are shown with an "lbl" prefix. The label "lblComBit" is configured with a fixed single border and AutoSize equal true. There are 10 unanimated labels. These labels are used only to convey static information. The required unanimated labels with the associated captions are as follows.

Label1: Caption "AB MicroLogix - Analog .... , 1 analog output"
Label2: Caption "Analog ->"
Label3: Caption "Volts/ma ->" with fixed single border and AutoSize = True
Label4: Caption "Temp/Press ->" with fixed single border and AutoSize = True
Label5: Caption "Oustside Temp" with fixed single border and AutoSize = True
Label6: Caption "Crawlspace Temp"with fixed single border and AutoSize =True
Label7: Caption "First Floor Temp" with fixed single border and AutoSize = True
Label8: Caption "Second Floor Temp" fixed single border and AutoSize = True
Label9: Caption "Great Room Temp" fixed single border and AutoSize = True
Label10: Caption "Water Pressure" with fixed single border and AutoSize = True

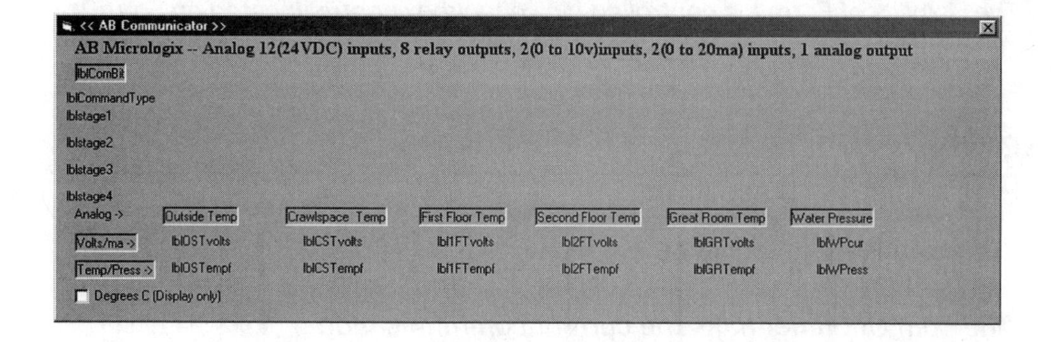

Figure 7.02

The required code for Form1 is as follows. For supplemental information on some of the sub procedures listed below, see the code presented in Chapter 5.

```
Option Explicit
Private Sub Form_Load()
' Set up the Communications Port
  MSComm1.CommPort = 2 'change according to your system configuration
   ' 9600 baud, no parity, 8 data, and 1 stop bit.
   MSComm1.Settings = "9600,N,8,1"
   ' Tell the control to read entire buffer when Input is used.
   MSComm1.InputLen = 1
   MSComm1.RThreshold = 1 'Set the characters to return to 1
   ' Open the port.
   MSComm1.PortOpen = True
  'set up PLC Communication Parameters
   bDST = 1 'destination address
   bSRC = 0 'source address: The computer is address zero
   READtimer.Interval = 75 ' Every 75 Millisecond the timer fires
   READtimer.Enabled = True
   ComTimeout.Enabled = False
   bReadNumberOfRegisters = 13
  lblComBit.Caption = " ComBit " 'eliminate the lbl prefix from the caption

 Form1.Visible = False 'Form1 is only displayed for calibration of the
temperature probes or for troubleshooting a communication problem.
The visibility of Form1 is controlled by a checkbox control located on Form2.

  Form2.Visible = True

End Sub
Private Sub ComTimeout_Timer()


'The Com Out timer halts the current communication
'transaction after a specified period of time.
'In this case the time limit is 250 milliseconds
```

```
bComStep = 0
ReadModeFlag = False
ClearAllcomflags
End Sub
```

```
Private Sub ClearAllcomflags()
'clears all communication flags and pointers

bLastLastRX = 0
bLastRX = 0
bNewRx = 0
STXflag = False
ETXflag = False
bRXpointer = 1
bETXposition = 0
End Sub
```

The READtimer sub procedure uses the ReadModeflag variable to alternate between a read command and a write command. The transaction number is incremented by one before each request. If for some reason the communication exchange takes longer than 75 milliseconds, the variable bComStep prevents another request from occurring before the existing request is satisfied or a communication time out is generated.

```
Private Sub READtimer_Timer()

If bComStep = 0 Then
   bComStep = 1 'Com Step 1
   clearhooks
   'This path if a message transaction is not going on
   'Increment the transaction number
  If lTransaction >= 65535 Then
   lTransaction = 0
   Else: lTransaction = lTransaction + 1
  End If
   'Alternate between read and write instructions.
```

```
'The first instruction is a read in order to gather the
'status of the miscellaneous write bits O/1, O/2, O/3

 If ReadModeFlag = False Then
   ReadModeFlag = True
   lblCommandType.Caption = "Read Command"
   GoRead
 Else: ReadModeFlag = False
   GOWRITE
  lblCommandType.Caption = "Write Command"
 End If
End If

End Sub
```

Labels are provided on Form1 to show the progress of the communication request. The communication protocol consists of four steps listed in the clearhooks Sub procedure.

```
Public Sub clearhooks()
'sets the text for the four stages of communication to the PLC
lblstage1.Caption = "Stage1-txt Command"
lblstage2.Caption = "Stage2- rx ack"
lblstage3.Caption = "Stage3-rx response"
lblstage4.Caption = "Stage4-Txt ack ...done"
End Sub
```

```
Private Sub GoRead()
  ' Load up the Read array
  Dim G As Byte
  bCMD = 1
  G = 1
  bTxData(G) = bDST  'Destination address (PLC =1)
  G = G + 1
   bTxData(G) = bSRC 'Source address (PC=0)
  G = G + 1
```

```
    bTxData(G) = bCMD 'Read or Write command
    G = G + 1
    bTxData(G) = STS 'always equal to zero
    G = G + 1
    bTxData(G) = (lTransaction And 255)
   ' lTRANSACTION is incremented before every message transmission
   ' when lTRANSACTION exceeds FFFF HEX it is cleared to 0000
    G = G + 1
    bTxData(G) = Int(lTransaction / 256)
    G = G + 1
    bTxData(G) = 0 '  address low byte N7:0
    G = G + 1
    bTxData(G) = 0  '  address high byte N7:0
    G = G + 1
    bTxData(G) = (bReadNumberOfRegisters * 2) 'size
    'place the end of data pointer in the array zero location
     bTxData(0) = G
ClearAllcomflags
    COMPUTECRC ' computes the crc based on data contained in bTxData array.
    '  LCRC is low byte of the result. bHCRC is the high byte of the result.
    GOTRANSMIT
End Sub
```

```
Private Sub GOWRITE()
    Dim G As Byte 'G is a scratch pad byte that serves as a counter of
                ' bytes entered into the transmit array
    bCMD = 8 'IN Allen-Bradley Protocol an Unprotected Write command is = to 8.
    G = 1
    bTxData(G) = bDST  'Destination address
    G = G + 1
    bTxData(G) = bSRC 'Source address
    G = G + 1
    bTxData(G) = bCMD 'WRITE
    G = G + 1
    bTxData(G) = STS ' Status Byte Constant
```

```
    G = G + 1
    bTxData(G) = (lTransaction And 255)
    ' lTRANSACTION is incremented before every message transmission
    ' when lTRANSACTION exceeds FFFF HEX it is cleared to 0000
    G = G + 1
    bTxData(G) = Int(lTransaction / 256) 'use Int to prevent rounding
    G = G + 1
    bTxData(G) = 100 ' write address low byte N7:50
    G = G + 1
    bTxData(G) = 0 ' write address high byte
    G = G + 1
    SetWriteBits
    bTxData(G) = bDATA  'low byte  represents outputs 0 through 7
    G = G + 1
    bTxData(G) = bDATA2
    bTxData(0) = G 'Place the total count in array location zero
  ClearAllcomflags
    COMPUTECRC ' computes the crc based on data contained in bTxData array.
    '  bLCRC is low byte of the result. bHCRC is the high byte of the result.

    GOTRANSMIT ' Sends the message out of the Communication Port.
  End Sub
```

The SetWriteBits Sub procedure is used to write to the outputs 0/1, 0/2, 0/3, the horn (0/0), and the daily register clear bit. The daily register clear bit will be sent at midnight by the animator program to clear the information contained in the daily registers. A Boolean AND operation is performed to clear only the bit being operated on. Subsequently, the bit being operated is set high, depending upon the state of the corresponding Visual Basic bit. In addition, the write ComBit is routed to read Combit. This routing of the ComBit keeps the ComBit dynamic.

```
  Public Sub SetWriteBits()

    Dim T As Byte
    If HornCmdFlag = True Then T = 1 Else T = 0
    bDATA = bDATA And 254 'Clear the bit to zero
```

```
  bDATA = bDATA Or T 'set bit to the value of T

   If Out1CmdFlag = True Then T = 2 Else T = 0
  bDATA = bDATA And 253 'Clear the bit to zero
  bDATA = bDATA Or T 'set bit to the value of T

 If Out2CmdFlag = True Then T = 4 Else T = 0
  bDATA = bDATA And 251 'Clear the bit to zero
  bDATA = bDATA Or T 'set bit to the value of T

 If Out3CmdFlag = True Then T = 8 Else T = 0
  bDATA = bDATA And 247 'Clear the bit to zero
  bDATA = bDATA Or T 'set bit to the value of T

 If ClearDailyRegCmdFlag = True Then T = 16 Else T = 0
  bDATA = bDATA And 239 'Clear the bit to zero
  bDATA = bDATA Or T 'set bit to the value of T

  If ComBitflag = True Then T = 128 Else T = 0
  bDATA2 = bDATA2 And 127 'Place the read combit into the write combit
    'first zero out bit 7 of the byte. Ultimately bit 15
        'Then OR with T which is the Read Combit
   bDATA2 = bDATA2 Or T 'set bit to the value of T
End Sub
```

```
Public Sub GOTRANSMIT()
txack 'transmit an acknowledge to close out any open transactions
vTxtData = "Transmitted Message => " 'initialize transmit readout
vRxData = "Received Message => " 'initialize receive readout
bLastLastRX = 0 'initialize receive history for ACK/NAK response
bLastRX = 0
bNewRx = 0
 ' Start of Messaging Packet
Let MSComm1.Output = Chr(DLE)
'Chr() Returns a String containing the character associated
```

```
'with the specified character code
vTxtData = vTxtData + CStr(Hex(DLE)) + ">"
'tack each transmitted character on to label caption with a ">" delimiter
Let MSComm1.Output = Chr(STX)
vTxtData = vTxtData + CStr(Hex(STX)) + ">"

Dim TXTpointer As Byte
 ' Start of APP Layer (All DLE's should be doubled up)
For TXTpointer = 1 To bTxData(0) ' bTxData(0) contains size of APP Layer

Let MSComm1.Output = Chr(bTxData(TXTpointer))
vTxtData = vTxtData + CStr(Hex(bTxData(TXTpointer))) + ">"
 'If the transmitted byte is a DLE, transmitted a second DLE
 If bTxData(TXTpointer) = DLE Then
   Let MSComm1.Output = Chr(DLE)
   vTxtData = vTxtData + CStr(Hex(DLE)) + ">"
 End If
Next TXTpointer

' End of Messaging Packet
Let MSComm1.Output = Chr(DLE)
vTxtData = vTxtData + CStr(Hex(DLE)) + ">"
Let MSComm1.Output = Chr(ETX)
vTxtData = vTxtData + CStr(Hex(ETX)) + ">"
Let MSComm1.Output = Chr(bLCRC)
vTxtData = vTxtData + CStr(Hex(bLCRC)) + ">"
Let MSComm1.Output = Chr(bHCRC)
vTxtData = vTxtData + CStr(Hex(bHCRC)) + ">"
If bComStep = 1 Then lblstage1.Caption = vTxtData
 'Initialize the time-out timer
ComTimeout.Enabled = True
Dim s1 'clear receive buffer
Let s1 = MSComm1.Input
End Sub
```

See Chapter 5 for additional information on the Sub procedures MSComm1_OnComm and GATHERresponse.

```
Private Sub MSComm1_OnComm()
'Static s1 As String
Dim s1 As String

Select Case MSComm1.CommEvent

    Case comEvReceive   ' Received RThreshold # of chars.
      Let s1 = MSComm1.Input
      vRxData = vRxData + CStr(Hex(Asc(s1))) + ">"
      bLastLastRX = bLastRX
      bLastRX = bNewRx
      bNewRx = Asc(s1)

     If bComStep = 2 Then
      If STXflag = True Then
       GATHERresponse
     Else: If bLastLastRX <> DLE And bLastRX=DLE And bNewRx = STX Then _
       STXflag = True: bRXpointer = 1
      End If
     End If

End Select

 If bComStep = 1 Then 'Step 1 wait for Ack or Nak after issuing request
If STXflag = False And bLastRX = DLE And bNewRx = NAK Then GOTRANSMIT
If STXflag = False And bLastRX = DLE And bNewRx = ACK Then
  bComStep = bComStep + 1 'Step 2: The uPLC received the request
   'Clear receive history and wait for DLE STX
  lblstage2.Caption = "Rx____________Ack"
  bLastLastRX = 0
  bLastRX = 0
  bNewRx = 0
  End If
```

```
 End If

If bComStep = 0 And bLastRX = DLE And bNewRx = ENQ Then_
 txack 'Resend Ack the uPlc did not receive it at step 4
 bLastLastRX = 0
 bLastRX = 0
 bNewRx = 0
 End If
End Sub
```

```
Private Sub GATHERresponse()
 Dim DLEstripFlag As Boolean
bRxData(bRXpointer) = bNewRx
 'strip off any extra DLE's
If bRXpointer > 1 And ETXflag=False And bNewRx = DLE And bLastRX = DLE_
Then bRXpointer=bRXpointer - 1: DLEstripFlag =True: bNewRx=0: bLastRX=0

'the App layer starts with   [DST][SRC][CMD][STS][TNS1][TNS2]
'bRXpointer settings          1 2   3  4   5   6
'no point in looking for an ETX until the pointer is > 6
If bRXpointer > 6 And ETXflag = False And bNewRx = ETX _
 And bRxData(bRXpointer - 1) = DLE And DLEstripFlag = False Then ETXflag=True:
bETXposition = bRXpointer

'Determine if the CRC bytes have been collected
If ETXflag = True And bRXpointer = bETXposition + 2 Then
    bRxData(0) = bETXposition - 2
   bRxHCRC = bRxData(bRXpointer)
   bRxLCRC = bRxData(bRXpointer - 1)
    Dim T As Byte
     For T = 0 To bRxData(0)
     bTxData(T) = bRxData(T)
     Next T
    COMPUTECRC
```

```
    If bRxHCRC <> bHCRC Or bRxLCRC <> bLCRC Then
   txNAK
   ClearAllcomflags
 End If

If bRxHCRC = bHCRC And bRxLCRC = bLCRC Then 'Received Good Message

  bComStep = bComStep + 1
  lblstage3.Caption = vRxData 'Good CRC:received a response
  txack 'Acknowledge the response
  ComTimeout.Enabled = False 'Stop Time-Out Timer
  bComStep = bComStep + 1
  'Communication transaction complete
  lblstage4.Caption = "Ack..................done!!!!!!"
  bComStep = 0

   If ReadModeFlag = True Then

     'After supplemental checks of proper size, proper destination address
         'and
     'transaction number agreement then collect the data and move to a
         'normalized
     'array
    If (bETXposition - 6) = 28 Then
  If bRxData(1)=0 And bRxData(5)=(lTransaction And 255) And bRxData(6) =_
Int(lTransaction / 256) Then

  'Move the PLC data to a byte array called "bPlcData". The data contained
  'in this array will be converted into the proper form by the sub procedure
  ' NormalizeData.

  For T = 7 To (bETXposition - 2)
      bPlcData(T - 6) = bRxData(T)
      Next T
      bPlcData(0) = (bETXposition - 2) - 6
      NormalizeData
```

```
        'Animate the ComBit
        If ComBitflag = True Then lblComBit.BackColor = QBColor(10) Else_
lblComBit.BackColor = QBColor(14)
          End If
        End If
       End If
  End If
End If

bRXpointer = bRXpointer + 1
End Sub
```

```
Private Sub txack()
Let MSComm1.Output = Chr(DLE)
Let MSComm1.Output = Chr(ACK)

End Sub
```

```
Private Sub txNAK()
Let MSComm1.Output = Chr(DLE)
Let MSComm1.Output = Chr(NAK)

End Sub
```

The following Sub procedure, NormalizeData, converts the data sent from the PLC into Visual Basic variables. Boolean logic is used to break out the bits contained in the first read register, N7:0. This register is now contained in two bytes located in the array bPlcData(). The first byte bPlcData(1) is the least significant byte of N7:0. The second byte bPlcData(2) is the most significant byte of N7:0. A binary progression is used to extract each byte and the corresponding Visual Basic bit or flag is set accordingly.

The first request of the PLC is a read. Only during the first acquisition, the status of the output bits O/1, O/2, and O/3 are routed to set the corresponding write bit. This initialization ensures that the status of the actual PLC outputs and the Visual

Basic outputs match. The read registers are mathematically pulled back together by multiplying the most significant byte by 256 and then adding the least significant byte. Variables dimensioned as Long are used to contain the result.

With respect to the temperature probes, an analog-to-digital value from zero to 31207 is produced with voltage input from zero to 10 volts. The voltage at the input is determined as follows: volts are equal to (A/D value) divided by 31207 times 10. If the A/D value is 31207, then the voltage is 31207/31207 * 10 = 1 * 10. If the A/D value is 15604, then the voltage is 15605/31207 * 10 or 5 volts.

The corresponding temperature in microamps is this voltage divided by the input impedance of 36.5 kohms in parallel with the 210 ohms of the analog input. An equivalent resistance can be determined by the following formula: Req=(R1*R2)/(R1+R2). The equivalent resistance calculates out to 31095 ohms. This resistance can also be measured with an ohmmeter if the PLC is not energized. With an input of 10 volts, the corresponding current is 10/31095, or 321.59 microamps. This is the current produced by the AD590 temperature-controlled, current source-integrated circuit. The AD590 emits one microamp per degree kelvin. Therefore, a measured voltage of 10 volts represents 321.59 degrees kelvin. Degrees kelvin is converted into degrees centigrade by subtracting a value of 273.15. Zero degree centigrade is 273.15 degrees kelvin. A temperature of 321.59 degrees kelvin is 48.44 degrees centigrade. The maximum temperature that can be measured with this configuration is 48.44 degrees centigrade, or 119.192 degrees Fahrenheit. Degrees Fahrenheit is converted to degrees centigrade by multiplying by 1.8 and adding 32.

Although the analog-voltage inputs of the Micrologix PLC adhere to an industrial standard of zero to 10 volts, the Micrologix PLC will actually read a maximum input voltage of 10.5 volts. An input of 10.5 volts produces an analog-to-digital value of 32767. The true maximum readable temperature is 148.14 degrees Fahrenheit.

Although the AD590 is trimmed at the factory, a calibration error called scale-factor error exists. The manufacturer proposes that the most elementary way to accomplish this calibration is to adjust the voltage-dropping resistor until the voltage output at the calibration temperature is equal to 1 millivolt per degree kelvin. Once this error is trimmed out, the remaining errors occur mostly at the temperature extremes of the device—at –55 degrees centigrade and at 150 degrees centigrade. A two-point trim could be used to further decrease the error. For this application, the worst-case range is the outside temperature sensor (a span of maybe 60 degrees centigrade), so a single-point trim should be sufficient.

In this project, the trim will be accomplished mathematically by multiplying the analog-to-digital determined input voltage by a span value. The span value is the ratio of the ideal voltage at calibration temperature divided by the actual measured voltage at calibration temperature. The ideal voltage is the voltage developed by the current of an ideal AD590 at the calibration temperature across the measured value of resistance, or in this case, 31100 ohms.

As an example, a high-quality laboratory thermometer with a range of 30 degrees centigrade and a resolution of 0.1 degree centigrade indicates a temperature of 11.3 degrees centigrade. The analog-to-digital value at this temperature, with the probe under calibration, is 27688. The corresponding voltage is 27688/31207 * 10, or 8.872368 volts. The ideal current is determined by converting the calibration temperature to Kelvin (273.15 + 11.3), or 284.45 microamps. This ideal current should develop a voltage of 284.45 microamps * 31100 ohms, or 8.846395 volts. The required span value is, therefore, 8.846395 / 8.872368. It is recommended that the span value be used in this form and not reduced to a value of 0.9970725.

The required temperature equation is (A/D Value/31207)*10* Span. Span is equal to the ratio of the ideal voltage at the calibration temperature divided by the calibration-measured voltage. The calibration-measured voltage is equal to the A/D value obtained at the calibration temperature divided by 31207 and multiplied by 10.

A centigrade display-only checkbox is provided to ease the calibration task. Working in centigrade reduces the amount of mathematical calculation required by the calibration task.

Temperature calibration is not as straightforward as you might assume. When you are performing a calibration, make certain that you select a point where the temperature probe and calibration device are stable—just like when you take your temperature during a fever. Allow time for both devices to reach a common temperature.

The pressure sensor used for this application produces a 4- to 20-milliamp signal for a pressure range of zero to 100 psi. The Micrologix analog-current input range is zero to 20 milliamps. The actual range of the sensor is 16 milliamps at 100 psi. Four milliamps is an offset value added to the signal. At a pressure of 50 psi, the sensor-output current is 12 milliamps. The analog-current input of the Micrologix sees a ratio of 12/20, or 60%. The pressure signal needs to be modified to produce the correct ratio. This modification is accomplished mathematically as follows.

```
Current =(Analog to digital value) / 31207 * 0.02  {.02 is equal to 20 milliamps}
Pressure = ((Current - 0.004) / 0.016) * 100 PSI
```

Now, a current of 12 milliamps produces the proper ratio of .5, and the pressure is correctly reported as 50 psi.

```
Public Sub NormalizeData()

'Break out the status bits
If (bPlcData(1) And 1) = 1 Then FrontDoorBellflag = True Else FrontDoorBellflag = False
If (bPlcData(1) And 2) = 2 Then RearDoorBellflag = True Else RearDoorBellflag = False
If (bPlcData(1) And 4) = 4 Then FrontDoorOpenflag = True Else FrontDoorOpenflag = False
If (bPlcData(1) And 8) = 8 Then RearDoorOpenflag = True Else RearDoorOpenflag = False
If (bPlcData(1) And 16) = 16 Then MailBoxflag = True Else MailBoxflag = False
If (bPlcData(1) And 32) = 32 Then FirstFloorHvacOnflag = True Else FirstFloorHvacOnflag = False
If (bPlcData(1) And 64) = 64 Then SecondFloorHvacOnflag = True Else
SecondFloorHvacOnflag = False
If (bPlcData(1) And 128) = 128 Then WaterPumpOnflag = True Else WaterPumpOnflag = False
If (bPlcData(2) And 128) = 128 Then ComBitflag = True Else ComBitflag = False
If (bPlcData(2) And 64) = 64 Then Status03flag = True Else Status03flag = False
If (bPlcData(2) And 32) = 32 Then Status02flag = True Else Status02flag = False
If (bPlcData(2) And 16) = 16 Then Status01flag = True Else Status01flag = False

'set 0/1, 0/2, 0/3 write bits to values contained in the PLC
'Do this only at program start-up

If StartUpflag = False Then
   StartUpflag = True ' set the flag true
   Out1CmdFlag = Status01flag
   Out2CmdFlag = Status02flag
   Out3CmdFlag = Status03flag
```

```
End If
'Collect analog and other remaining data

'Convert the transmitted bytes back to the respective registers

 lRawOutsideTemp = (bPlcData(4) * 256) + bPlcData(3)
 lRawCrawlSpaceTemp = (bPlcData(6) * 256) + bPlcData(5)
 lRawFirstFloorTemp = (bPlcData(8) * 256) + bPlcData(7)
 lRawSecondFloorTemp = (bPlcData(10) * 256) + bPlcData(9)
 lRawGreatRoomTemp = (bPlcData(12) * 256) + bPlcData(11)
 lRawWaterPressure = (bPlcData(14) * 256) + bPlcData(13)
 lHvac1DailyCycle = (bPlcData(16) * 256) + bPlcData(15)
 lHvac1DailyOnTime = (bPlcData(18) * 256) + bPlcData(17)
 lHvac2DailyCycle = (bPlcData(20) * 256) + bPlcData(19)
 lHvac2DailyOnTime = (bPlcData(22) * 256) + bPlcData(21)
 lWaterPumpDailyCycle = (bPlcData(24) * 256) + bPlcData(23)
 lWaterPumpDailyOnTime = (bPlcData(26) * 256) + bPlcData(25)
 'Temporary scratch pad variables
 Dim Volts
 Dim Current
 Dim DegreeC

 ' Normalize Data: Convert into units

'The Allen-Bradley 10 Volt analog input provides an analog to digital value of 31207
' when 10 volts is supplied to the input.  The form of the conversion equation is
' volts = A/D(0 to 31207) / 31207 * 10
'A single point span is used for calibration and changes the conversion equation as
'follows:
'volts = volts * (ideal voltage at known temperature with 31100 Ohm resistor)/Actual
'volts read.

 'The Format keyword is used limit the number of decimal places.

  Volts = lRawOutsideTemp / 31207 * 10 * 8.846395 / 8.872368
  lblOSTvolts.Caption = Format(Volts, "#0.000000")
```

```
 DegreeC = ((Volts / 31100 * 1000000) - 273.15)
 vOutsideTemp = DegreeC * 1.8 + 32
  vOutsideTemp = Format(vOutsideTemp, "##0.000")
If cbDisplayC.Value = 0 Then
 lblOSTempf.Caption = vOutsideTemp
 Else: lblOSTempf.Caption = Format(DegreeC, "##0.0000")
End If

Volts = lRawCrawlSpaceTemp / 31207 * 10 * 8.9332365 / 8.9246
 lblCSTvolts.Caption = Format(Volts, "#0.000000")
 DegreeC = ((Volts / 31100 * 1000000) - 273.15)
 vCrawlSpaceTemp = DegreeC * 1.8 + 32
  vCrawlSpaceTemp = Format(vCrawlSpaceTemp, "##0.000")
If cbDisplayC.Value = 0 Then
 lblCSTempf.Caption = vCrawlSpaceTemp
 Else: lblCSTempf.Caption = Format(DegreeC, "##0.0000")
End If

Volts = lRawFirstFloorTemp / 31207 * 10 * 8.950 / 8.92
 lbl1FTvolts.Caption = Format(Volts, "#0.000000")
 DegreeC = ((Volts / 31100 * 1000000) - 273.15)
 vFirstFloorTemp = DegreeC * 1.8 + 32
  vFirstFloorTemp = Format(vFirstFloorTemp, "##0.000")
 If cbDisplayC.Value = 0 Then
  lbl1FTempf.Caption = vFirstFloorTemp
  Else: lbl1FTempf.Caption = Format(DegreeC, "##0.0000")
 End If

Volts = lRawSecondFloorTemp / 31207 * 10 * 8.962 / 8.971
lbl2FTvolts.Caption = Format(Volts, "#0.000000")
DegreeC = ((Volts / 31100 * 1000000) - 273.15)
 vSecondFloorTemp = DegreeC * 1.8 + 32
  vSecondFloorTemp = Format(vSecondFloorTemp, "##0.000")
If cbDisplayC.Value = 0 Then
```

```
 lbl2FTempf.Caption = vSecondFloorTemp
 Else: lbl2FTempf.Caption = Format(DegreeC, "##0.0000")
End If

Volts = lRawGreatRoomTemp / 31207 * 10 * 8.97 / 8.9523
lblGRTvolts.Caption = Format(Volts, "#0.000000")
DegreeC = ((Volts / 31100 * 1000000) - 273.15)
 vGreatRoomTemp = DegreeC * 1.8 + 32
   vGreatRoomTemp = Format(vGreatRoomTemp, "##0.000")
If cbDisplayC.Value = 0 Then
 lblGRTempf.Caption = vGreatRoomTemp
 Else: lblGRTempf.Caption = Format(DegreeC, "##0.0000")
End If


Current = lRawWaterPressure / 31207 * 0.02
lblWPcur.Caption = Format((Current * 1000), "#0.000") 'convert to milliamps
 vWaterPressure = ((Current - 0.004) / 0.016) * 100
  vWaterPressure = Format(vWaterPressure, "##0.000")
   lblWPress.Caption = vWaterPressure

'send data to a file for access by other remote users
CreateDataFile

End Sub
```

The CreateDataFile Sub procedure saves all the bits and normalized register values in an ASCII file called "HomeMonitor.csv". This file can then be accessed remotely by other computers in read-only mode. Essentially, this form would be modified at the remote computer to load the data from the file into the respective Visual Basic variables. The animation forms use the Visual Basic variables to perform the animation tasks.

```
Private Sub CreateDataFile()
        Dim AllData

            AllData = CStr(FrontDoorBellflag) + "," _
```

```
        + CStr(RearDoorBellflag) + "," _
        + CStr(FrontDoorOpenflag) + "," _
        + CStr(RearDoorBellflag) + "," _
        + CStr(MailBoxflag) + "," _
        + CStr(FirstFloorHvacOnflag) + "," _
        + CStr(SecondFloorHvacOnflag) + "," _
        + CStr(WaterPumpOnflag) + "," _
        + CStr(vOutsideTemp) + "," _
        + CStr(vCrawlSpaceTemp) + "," _
        + CStr(vFirstFloorTemp) + "," _
        + CStr(vSecondFloorTemp) + "," _
        + CStr(vGreatRoomTemp) + "," _
        + CStr(vWaterPressure) + "," _
        + CStr(lHvac1DailyCycle) + "," _
        + CStr(lHvac1DailyOnTime) + "," _
        + CStr(lHvac2DailyCycle) + "," _
        + CStr(lHvac2DailyOnTime) + "," _
        + CStr(lWaterPumpDailyCycle) + "," _
        + CStr(lWaterPumpDailyOnTime) + "," _
        + CStr(ComBitflag)
    On Error Resume Next
    SetAttr "HomeMonitor.csv", vbNormal
    Dim Filenum
    Filenum = FreeFile
    Open "HomeMonitor.csv" For Output As #Filenum
    Print #Filenum, AllData
    Close #Filenum
    SetAttr "HomeMonitor.csv", vbReadOnly

End Sub
```

A Form_QueryUnload Sub procedure is provided to interrupt a Form1 close statement. This Sub procedure makes Form1 invisible and sets the initiating checkbox "cbShowCommunicationWin" located on Form2 accordingly.

```
Private Sub Form_QueryUnload(Cancel As Integer, UnloadMode As Integer)
```

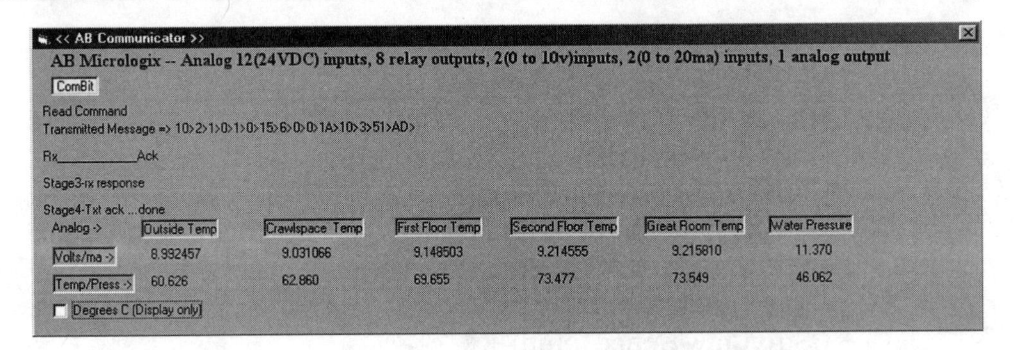

Figure 7.03

```
'if the visible form is told to close then make the form not visible
'Cancel the close command and change the state of the checkbox on form2
Form1.Visible = False
Cancel = True
Form2.cbShowCommunicationWin.Value = False 'A checkbox is provided
on form2 to 'make form1 visible or not visible
End Sub
```

An actual snapshot of a running form is provided in Figure 7.03.

Conclusion/Exercise

As an exercise, you should modify the calibration process to allow the calibration temperature for each temperature probe to be keyed in with a textbox. Once the calibration value is entered, write the required code to automatically calculate the span value. This feature requires the span values to be saved to a file and retrieved from the file on program start-up.

Chapter 8
HOME-MONITOR ANIMATION MODULE

INTRODUCTION

The acquisition module covered in Chapter 7 acquired and assigned various PLC bits and registers to Visual Basic variables. This chapter animates a form with the acquired data. Material discussed in all the chapters from this point on works with the information acquired by Form1. In this example, it's an Allen-Bradley Micrologix PLC, but it could be any other PLC, serial device, or combination of PLCs and serial devices.

Task Description

Add a form to the project called Form2 and change the caption to read "Home Monitor". The proposed layout of Form2 is presented in Figure 8.01. The form contains two timers, a label for the Combit, a checkbox to control the visibility of Form1, a label for displaying real-time information, labels to set/show the status of the available controllable outputs of the PLC, and various frame controls. Two timers are required. The first timer is called "tmrUpdate". This timer is activated every 50 milliseconds and is used to provide the animation. All form controls are updated with each "tmrUpdate" event. The second timer is called "tmrWaterPump". This timer is used to provide motion to two pump shapes whenever the pump is running. Unless otherwise specified, all the labels on Form2 are configured as AutoSize=True and BorderStyle="1-Fixed Single". The label font sizes are 10, 12, and 18. The darker fonts are in bold.

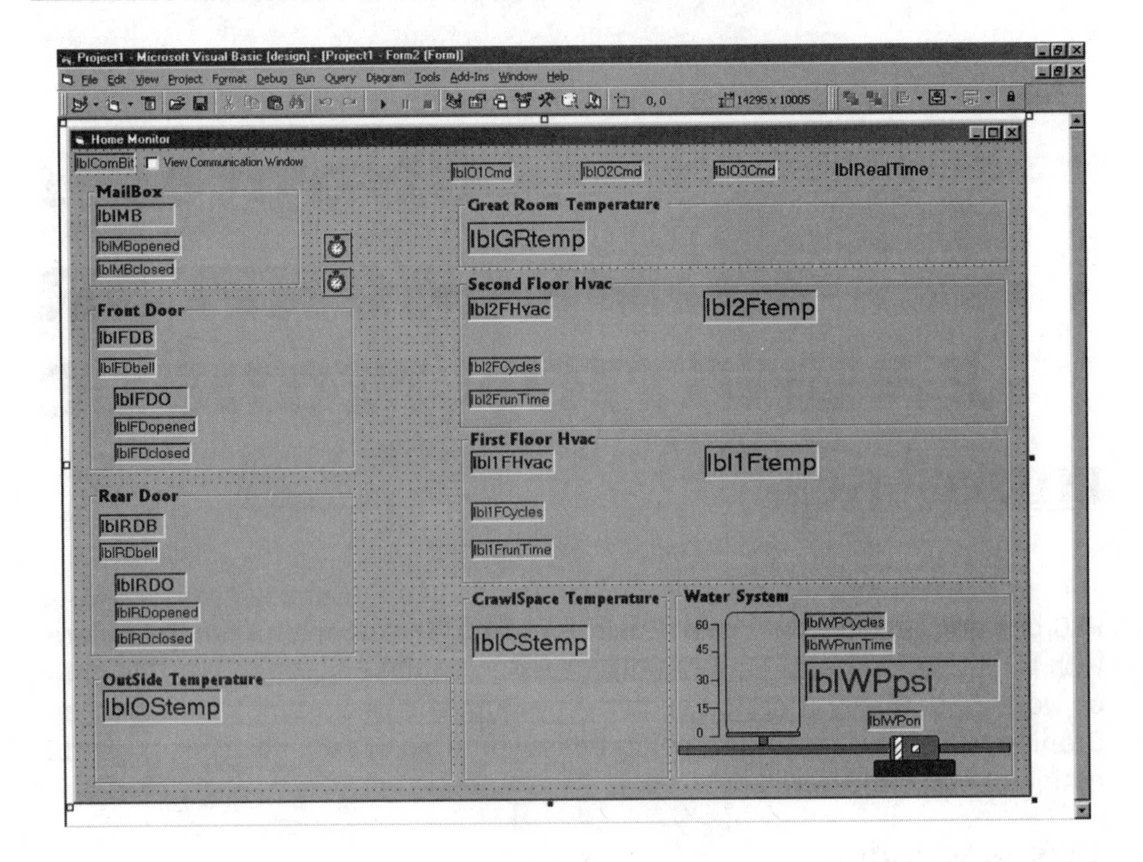

Figure 8.01

The Combit label will alternate between a background of green and a background of yellow. This alternation is used to provide a visual indication that the PC is successfully communicating with the PLC.

A CheckBox control called "cbShowCommunicationWin" is used to control the visibility property of Form1. This feature will come in handy when the temperature probes require calibration.

A label called "lblRealTime" provides a display of the present time and date. The BorderStlye attribute for this label is "0-none".

Three labels are provided to show the status of the spare outputs of the PLC. The spare or unassigned PLC outputs are O/1, O/2, and O/3. These labels provide true status of the outputs as transmitted from the PLC. In addition, if any of the labels are double-clicked, the corresponding PLC output will change state and an audio beep is generated.

Frame 1 is used to contain three labels for the Mailbox Status. Label "lblMB" is used to indicate whether the mailbox door is opened or closed. The background color of the label is changed to red whenever the mailbox door is open. Label "lblMBopened" provides time and date information of the last mailbox-open event. Label "lblMBclosed" provides time and date information on when the mailbox door was closed. Whenever the mailbox door is opened, the mailbox time/date closed caption is changed to " Closed: ?" until the door is closed.

Frame 2 is used to contain five labels for the Front Door Status. The label "lblFDB" is used to indicate the status of the front doorbell switch. This label will change from "Front Door Bell Switch Open" to "Front Door Bell Switch Closed" when the switch is pressed. In addition, the label background color will change to green when the switch is pushed. The label "lblFDbell" provides time/date information on when the doorbell switch was pressed. Label "lblFDO" is used to indicate whether the front door is opened or closed. The background color of the label is changed to red whenever the door is open. Label "lblFDopened" provides time and date information of the last door-open event. Label "lblFDclosed" provides the time and date information on when the door was closed. Whenever the door is opened, the front door time/date closed caption is changed to " Closed: ?" until the door is closed.

Frame 3 is used to contain five labels for the Rear Door Status. The associated labels function identically to the front door labels described in regards to Frame 2.

Frame 4 is used to show outside temperature in degrees Fahrenheit. Similarly, Frame 5 is used to show crawlspace temperature in degrees Fahrenheit.

Frame 6 contains four labels. Label "lbl1FHvac" shows the on/off status of the heating system. The number of cycles that the first floor heating system experienced is shown in the label "lbl1Fcycles". The total daily run time is displayed in the label "lbl1FrunTime". Label "lbl1Ftemp" shows the current temperature of the first floor in degrees Fahrenheit.

Frame 7 is identical to Frame 6, except the parameters reflect the status of the heating system controlling the second floor.

Frame 8 contains a single label called “lblGRtemp”. This label shows the current temperature of the great room in degrees Fahrenheit.

Frame 9 is used to display the status of the water-supply system. In this case, the water system is a well pump with a water-storage tank. The label “lblWPpsi” is used to show the pressure reading from the pressure transducer. The number of water-pump cycles is displayed by Label “lblWPCycles”. Total pump run time is indicated by label “lblWPrunTime”. A close-up view of the water system is provided in Figure 8.02.

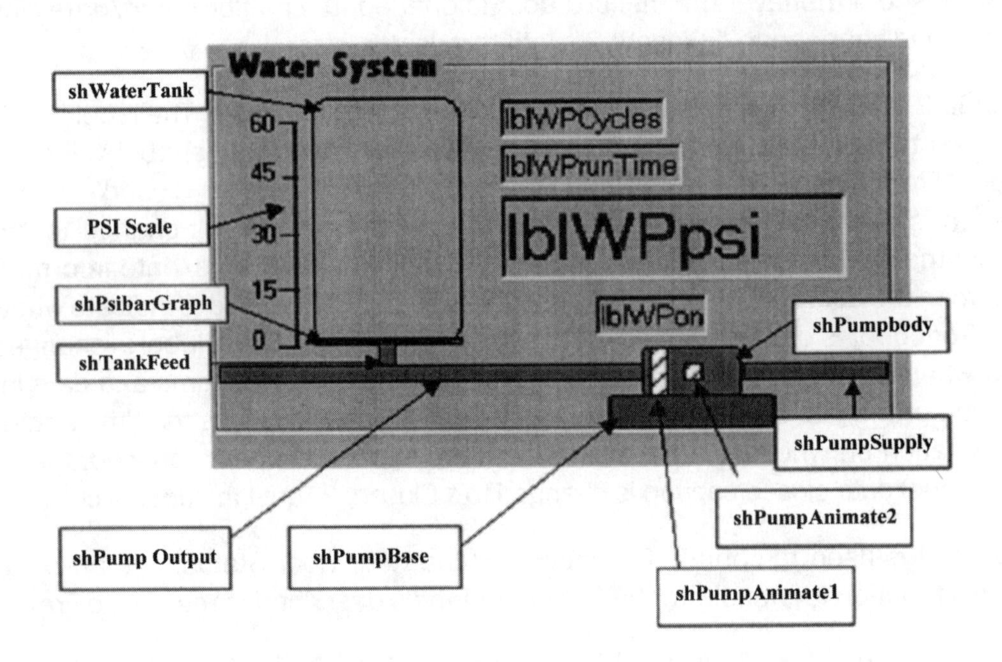

Figure 8.02

Several shapes are used to construct an overview of the key water-system components.

1) “shWaterTank” is a default shape with a shape type “4-Rounded Rectangle”. This shape represents the water tank.
2) “shPumpOutput” is a default rectangular shape with a fill color of blue and fill style of “0-Solid”. This shape represents the water-supply line.

3) "shTankFeed" is a default rectangular shape with a fill color of blue and fill style of "0-Solid". This shape represents the water feed to the tank.
4) "ShPumpSupply" is a default rectangular shape with a fill color of blue and fill style of "0-Solid". This shape represents the pump-supply water feed.
5) "ShPumpBase" is a default rectangular shape with a fill color of black and fill style of "0-Solid". This shape represents the pump base.
6) "ShPumpbody" is a default shape with a shape type "4-Rounded Rectangle", a fill style of "0-Solid" and a fill color of blue. This shape is used to represent the pump body.
7) "shPumpAnimate1" is a default rectangular shape with a fill style of "5-Downward Diagonal". This shape will move laterally whenever the pump is on. The setting for this shape must be "Bring to Front" with respect to the pump body.
8) "shPumpAnimate2" is a default rectangular shape with a fill style of "5-Downward Diagonal". This shape will move laterally whenever the pump is on. The setting for this shape must be "Bring to Front" with respect to the pump body.
9) "shPSIbarGraph" is a default rectangular shape with a fill color of blue and fill style of "0-Solid". The size of this shape will change proportionally with system pressure.
10) The pressure scale is created using lines and labels. A description of the scale creation is located later in this chapter.

Pump motion is created using the shapes "shPumpAnimate1" and "shPumpAnimate2". This simple example is driven by the state of the water-pump run status and the timer "tmrWaterPump". Timer "tmrWaterPump" has an interval set to one millisecond. The animation of both shapes is coordinated by this time. A private static integer variable called "X" is either incremented or decremented by 50 with every "tmrWaterPump" event. Initially, the value of "X" is added to the left position property of both pump-animation shapes. The addition causes both shapes to shift 50 twips to the right. A twip is a unit of length equal to 1/20 of a printer's point, and a printer's point is 1/72 of an inch. The twip-origination point is the upper left side of the controlling container. In this situation, the controlling container is Frame 9. When "X" reaches a value of 500, a private static

Boolean flag called "DownCountFlag" is set true, which causes the next timer event to subtract a value of 50 from the shape's left position property. The Boolean flag remains in the true state until "X" reaches a value of zero. This cycle continues until the pump-run status changes to off. At this time, both shapes are returned to the original starting location. The original starting location was stored in the variables "IS1Pos" and "IS2Pos" during the Form2 Load_Event. The code to perform this function is located in Sub procedure "tmrWaterPump_Timer".

The next animation task requires that the rectangular shape "ShPSIbarGraph" proportionally change size to graphically reflect the pressure level of the water tank. You should start with the creation of the required psi scale. The water system is designed to operate at a maximum psi of 60. The maximum level of the scale will then be 60, and gradients are placed at 25% increments. Labels are used to display the pressure value at each gradient. The required scale components are depicted in Figure 8.03.

Figure 8.04 details the twip location (x,y) for each line element. Each gradient level was mathematically determined based on the length of line 1. Line 1 represents the desired range of the water-pressure variable. The vertical length of line 1 is equal to the 2220 –570, or 1650 twips.

The tricky part in determining the location of each gradient is the location of the origin in the upper left-hand corner. Most people are familiar with the Cartesian coordinate numbering system, which places the origin in the bottom left-hand corner. In the Cartesian coordinate system, vertical numbers increase as you go vertical. With the origin located in the upper left-hand corner, the number decreases as you go vertical. This can be easily dealt with mathematically, but it's a little disorientating initially.

The 25% gradient level (or 15 psi point) is determined by multiplying the twip range of 1650 by .25 and then subtracting the result from the vertical value of the lower end of line 1 (or 2220). Therefore, horizontal line 3 is located at a vertical value of 1807. Similarly, the twip location of 30 psi is determined by multiplying the desired span in twips (or 1650) and subtracting the result from the y value of the lower end of line 1 (or 2220). This calculation yields a vertical value of 1395. The positions of finer gradient values are calculated in the same manner.

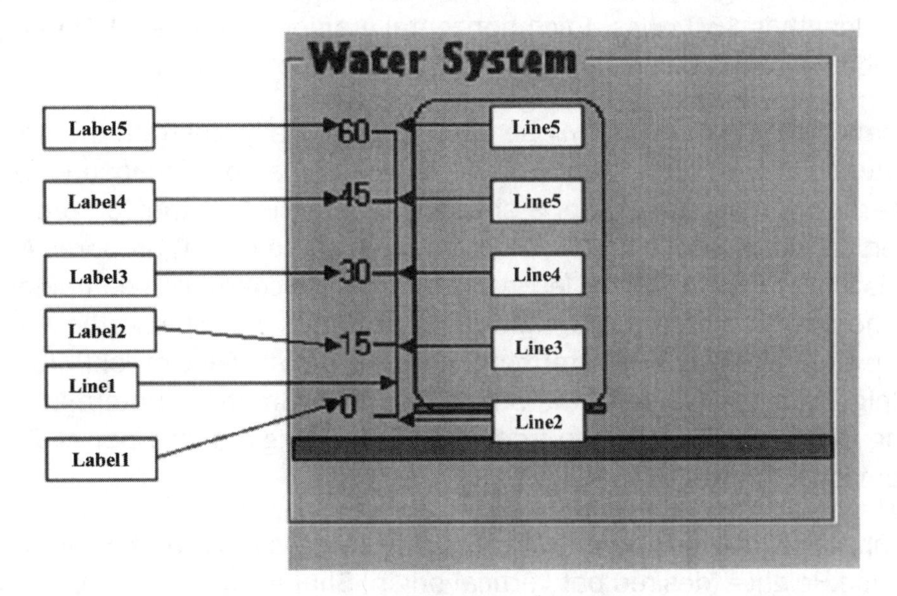

Figure 8.03

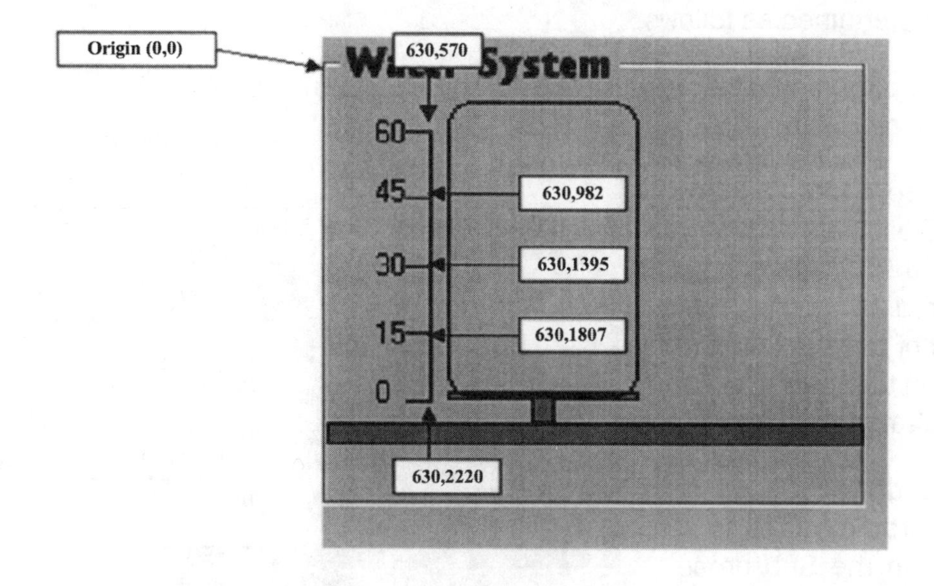

Figure 8.04

The horizontal length of lines 2 through 5 is a matter of preference. In this case, the line length is 150 twips. Each horizontal gradient line has a left end point at 500 and a right end point at 650. The y value doesn't change.

The shape "shPSIbarGraph" will be used to graphically illustrate the pressure of the water tank. The mathematics for this task is similar to psi measurement scale. The starting vertical position of a shape is determined by the Top property. The end vertical position of a shape is determined by the Height property. Again, the origin is located at the upper left hand side of the controlling container. To animate the pressure as a positive-growing bar graph, we will use the ratio of the actual pressure to 60 psi to determine the Top property value of "shPSIbarGraph". The Height property value of the shape "shPSIbarGraph" is determined by subtracting the Top property value from the vertical origin of the bar at 2220. The mathematical formula is as follows.

Shape.Top= (desired bar vertical origin)-(Actual Press./Full Press.*twip span)
Shape.Height= (desired bar vertical origin)-Shape.Top

In this case, use 45 psi as an example of an actual pressure reading. Shape top is then determined as follows.

Shape.Top = 2220- (45/60*1650) = 2220-1237 = 982
Shape.Height= 2220-982 = 1238

Therefore, the shape "shPSIbarGraph" will start at twip 982 and expand down 1238 twips. The width of the shape will remain in tact. This example is illustrated in Figure 8.05.

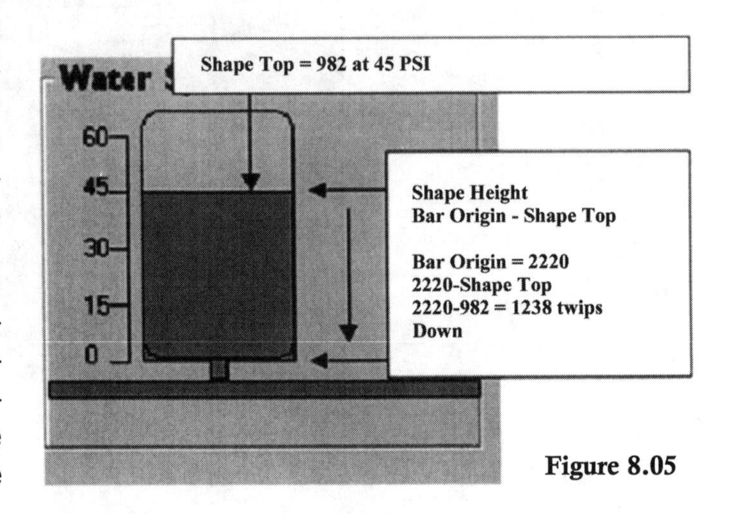

Figure 8.05

The code for the water-pressure bar graph is located in the Sub procedure ShowAnalog. A large portion of the time, the pressure will be stationary, so there is no need to replot the bar graph. An "If statement" is provided that excludes bar graph function if the pressure change is less than plus or minus 0.05 from the last pressure reading.

The Code

The following section of program code represents all the code required for the functions described. This code is provided on the companion CD-ROM. The code is written in sequential fashion and should be easy to follow. Note the use of the message box in the Form_Unload event, which provides an opportunity to cancel an accidental closing of the program. Program the following.

```
Option Explicit
Public lS1Pos As Long
Public lS2Pos As Long
_______________________________________________

Private Sub cbShowCommunicationWin_Click()
 If cbShowCommunicationWin.Value = 1 Then Form1.Visible = True Else
Form1.Visible = False
End Sub
_______________________________________________

Private Sub Form_Load()
'Get the positions of the pump animation shapes
 lS1Pos = shPumpAnimate1.Left
 lS2Pos = shPumpAnimate2.Left
'Set the following captions to eliminate the label identifiers
lblComBit.Caption = " ComBit "
lblO1Cmd.Caption = " O/1 Command "
lblO2Cmd.Caption = " O/2 Command "
lblO3Cmd.Caption = " O/3 Command "

End Sub
_______________________________________________

Private Sub tmrUpdate_Timer()
'Show real time
lblRealTime.Caption = Now
```

```
'At midnight send a Register clear command to the PLC
'for all timers and Cycle counters. Note although the flag is held high
'for approximately 10 seconds. The PLC conditions the flag with a
'one shot coil resulting in a reset of only 1 scan time duration.

If Hour(Now()) = "0" And Minute(Now()) = "0" And Second(Now()) < 10 Then
ClearDailyRegCmdFlag = True
Else: ClearDailyRegCmdFlag = False
End If

ShowFlags 'animate bits
ShowTimesAndCycles  'show run times and number of cycles
ShowAnalog 'show all analog data


End Sub
```

```
Public Sub ShowFlags()

 If ComBitflag = True Then lblComBit.BackColor = QBColor(10) Else
lblComBit.BackColor = QBColor(14)
 If Status01flag = True Then lbl01Cmd.BackColor = QBColor(14) Else
lbl01Cmd.BackColor = QBColor(7)
 If Status02flag = True Then lbl02Cmd.BackColor = QBColor(14) Else
lbl02Cmd.BackColor = QBColor(7)
 If Status03flag = True Then lbl03Cmd.BackColor = QBColor(14) Else
lbl03Cmd.BackColor = QBColor(7)


If FrontDoorBellflag = True Then
lblFDB.BackColor = QBColor(10)
lblFDB.Caption = " Front Door Bell Switch On  "
lblFDbell.Caption = "Bell Rung: " & Now
Else: lblFDB.BackColor = QBColor(7)
lblFDB.Caption = " Front Door Bell Switch Off "
End If

If RearDoorBellflag = True Then
```

```
lblRDB.BackColor = QBColor(10)
lblRDB.Caption = " Rear Door Bell Switch On  "
lblRDbell.Caption = "Bell Rung: " & Now
Else: lblRDB.BackColor = QBColor(7)
lblRDB.Caption = " Rear Door Bell Switch Off "
End If

If FrontDoorOpenflag = True Then
lblFDO.BackColor = QBColor(12)
lblFDO.Caption = "*     Front Door Open      *"
Else: lblFDO.BackColor = QBColor(7)
lblFDO.Caption = "     Front Door Closed     "
End If

If RearDoorOpenflag = True Then
lblRDO.BackColor = QBColor(12)
lblRDO.Caption = "*     Rear Door Open       *"
Else: lblRDO.BackColor = QBColor(7)
lblRDO.Caption = "     Rear Door Closed     "
End If

If MailBoxflag = True Then
lblMB.BackColor = QBColor(12)
lblMB.Caption = "*     Mail Box Open      *"
Else: lblMB.BackColor = QBColor(7)
lblMB.Caption = "     Mail Box Closed     "
End If

If FirstFloorHvacOnflag = True Then
lbl1FHvac.BackColor = QBColor(10)
lbl1FHvac.Caption = "   First Floor Hvac On   "
Else: lbl1FHvac.BackColor = QBColor(7)
lbl1FHvac.Caption = "   First Floor Hvac Off   "
End If

If SecondFloorHvacOnflag = True Then
```

```
lbl2FHvac.BackColor = QBColor(10)
lbl2FHvac.Caption = "  Second Floor Hvac On  "
Else: lbl2FHvac.BackColor = QBColor(7)
lbl2FHvac.Caption = "  Second Floor Hvac Off  "
End If

If WaterPumpOnflag = True Then
  lblWPon.BackColor = QBColor(10)
  lblWPon.Caption = " Water Pump On "
  tmrWaterPump.Enabled = True
Else: lblWPon.BackColor = QBColor(7)
  lblWPon.Caption = " Water Pump Off "
  tmrWaterPump.Enabled = False
  ' Place pump animation members back at the initial location
  shPumpAnimate1.Left = IS1Pos
  shPumpAnimate2.Left = IS2Pos
End If

CheckDoorChange

End Sub
```

```
Public Sub CheckDoorChange()

If FrontDoorOpenflag <> LastFrontDoorOpenflag Then
   LastFrontDoorOpenflag = FrontDoorOpenflag
  If FrontDoorOpenflag = True Then
     lblFDopened.Caption = " Opened: " & Now
     lblFDclosed.Caption = " Closed: ?"
  Else:
     lblFDclosed.Caption = " Closed: " & Now
  End If
 End If

 If RearDoorOpenflag <> LastRearDoorOpenflag Then
   LastRearDoorOpenflag = RearDoorOpenflag
```

```
    If RearDoorOpenflag = True Then
       lblRDopened.Caption = " Opened: " & Now
       lblRDclosed.Caption = " Closed: ?"
    Else:
       lblRDclosed.Caption = " Closed: " & Now
    End If
 End If

 If MailBoxflag <> LastMailBoxflag Then
    LastMailBoxflag = MailBoxflag
    If MailBoxflag = True Then
       lblMBopened.Caption = " Opened: " & Now
       lblMBclosed.Caption = " Closed: ?"
    Else:
       lblMBclosed.Caption = " Closed: " & Now
    End If
 End If

End Sub
```

```
Private Sub ShowTimesAndCycles()
'times and cycles

lblWPCycles.Caption = "Cycles Today => " + CStr(lWaterPumpDailyCycle)
lblWPrunTime.Caption = "RunTime Today: " + CStr(lWaterPumpDailyOnTime_
/ 10) + " Minutes "
lbl1FCycles.Caption = "Cycles Today => " + CStr(lHvac1DailyCycle)
lbl1FrunTime.Caption = "RunTime Today: " + CStr(lHvac1DailyOnTime) + _
" Minutes "
lbl2FCycles.Caption = "Cycles Today => " + CStr(lHvac2DailyCycle)
lbl2FrunTime.Caption = "RunTime Today: " + CStr(lHvac2DailyOnTime) +_
" Minutes "

End Sub
```

```
Private Sub ShowAnalog()
```

```
'Analog
lblWPpsi.Caption = " " + vWaterPressure + " PSI"
lblCStemp.Caption = " " + vCrawlSpaceTemp + " Deg F "
lblOStemp.Caption = " " + vOutsideTemp + " Deg F "
lbl1Ftemp.Caption = " " + vFirstFloorTemp + " Deg F "
lbl2Ftemp.Caption = " " + vSecondFloorTemp + " Deg F "
lblGRtemp.Caption = " " + vGreatRoomTemp + " Deg F "

Static vlastpsi ' Save the last PSI reading
'A large portion of the time the pressure will be stationary and there is no need to
'redo the bar graph. The following if statement performs the bar graph only if
'the pressure changed + or - .05 from the last reading
If vlastpsi = "" Then vlastpsi = 0
If vWaterPressure < 0 Then Let vWaterPressure = 0 'prevent error if negative
If (vWaterPressure - vlastpsi) > 0.05 Or (vWaterPressure - vlastpsi) < -0.05 Then
  shPSIbarGraph.Top = 2220 - ((vWaterPressure / 60) * 1650)
  shPSIbarGraph.Height = 2220 - shPSIbarGraph.Top
  vlastpsi = vWaterPressure
End If
End Sub
```

```
Private Sub tmrWaterPump_Timer()

Static x As Integer 'Local Scratch pad Variable
Static DownCountFlag As Boolean
If DownCountFlag = False Then
    x = x + 50 ' shift to the right
    'Add x to the original position of both shapes
    shPumpAnimate1.Left = lS1Pos + x
    shPumpAnimate2.Left = lS2Pos + x
    If x > 500 Then DownCountFlag = True
 Else: x = x - 50 ' shift to the left
    'Add x to the original position of both shapes
     shPumpAnimate1.Left = lS1Pos + x
     shPumpAnimate2.Left = lS2Pos + x
If x = 0 Then DownCountFlag = False
```

```
End If
End Sub
```

```
Private Sub lblO1Cmd_DblClick()
Out1CmdFlag = Out1CmdFlag Xor True
Beep
End Sub
```

```
Private Sub lblO2Cmd_DblClick()
Out2CmdFlag = Out2CmdFlag Xor True
Beep
End Sub
```

```
Private Sub lblO3Cmd_DblClick()
Out3CmdFlag = Out3CmdFlag Xor True
Beep
End Sub
```

```
Private Sub Form_Unload(Cancel As Integer)

'Use confirmation before shutting down
Dim command As Integer
'vbQuestion =>Display Warning Query icon.
command = MsgBox("Are you sure you want to exit this Program?", vbYesNo +
vbQuestion)
If command = vbYes Then
   MsgBox ("Adios, it's been real")
    Else: Cancel = True
End If

Unload Me
Unload Form1
End

End Sub
```

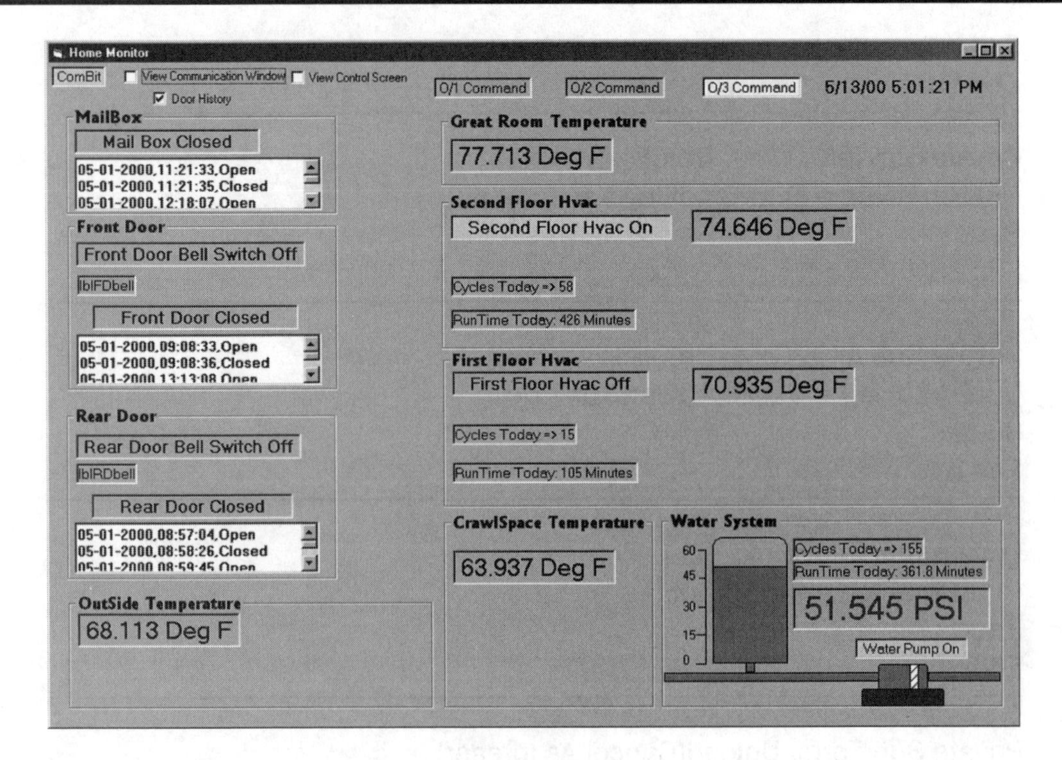

Figure 8.06

An actual running home-monitor Form is shown in Figure 8.06.

Conclusion/Exercise

As an exercise, you should enhance the animation program to include daily high and low temperatures, along with an associated time stamp for all temperature readings. These values should be saved to the hard drive and restored if the program should be halted. In addition, include any all-time high and low values for both outside temperature and crawlspace temperature. The all-time high and low readings should include time and date information.

Chapter 9
HOME ANIMATION: WAVE FILES

INTRODUCTION

In this chapter, you can have some fun and launch wave files for specific events. Now that you have the doorbell push buttons wired into the PLC, you need an audio annunciation. At the same time, you can provide audio sound for all the door events. So scan the Web for some interesting wave files to include in this project.

Task Description

The first component to add to the home-animation code is the Multimedia MCI control. This control manages the playback and recording of multimedia files on Media Control Interface (MCI) devices. MCI devices include audio cards, MIDI sequencers, CD-ROM players, and so on. The Multimedia MCI control is denoted in Visual Basic as MMcontrol. The control is packaged with a number of control buttons; however, the MMcontrol object can be controlled with the intrinsic buttons or with software.

Add the MMcontrol to the toolbar by right-clicking on the toolbar. Then left-click, selecting components. Under the control tab, scroll down and place a check in the "Microsoft Multimedia Control 6.0" checkbox. Figure 9.01 illustrates this task.

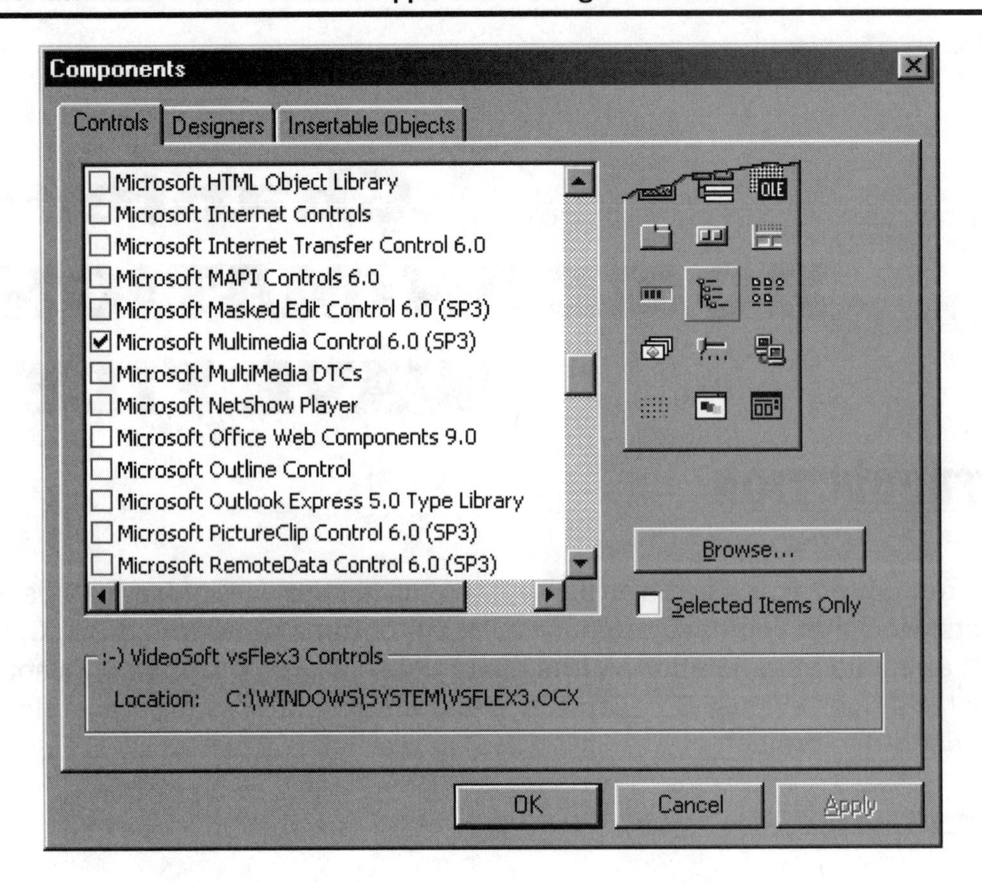

Figure 9.01

Add the control to Form2 directly under the lblRealTime, as illustrated in Figure 9.02. The multimedia control in this project will be strictly under software control, and the MMcontrol object's visible property will be specified as false.

The MMcontrol is very easy to use and requires only several run-time properties to be configured. These properties will be discussed as you progress with the code. Microsoft Help should be used to provide additional information.

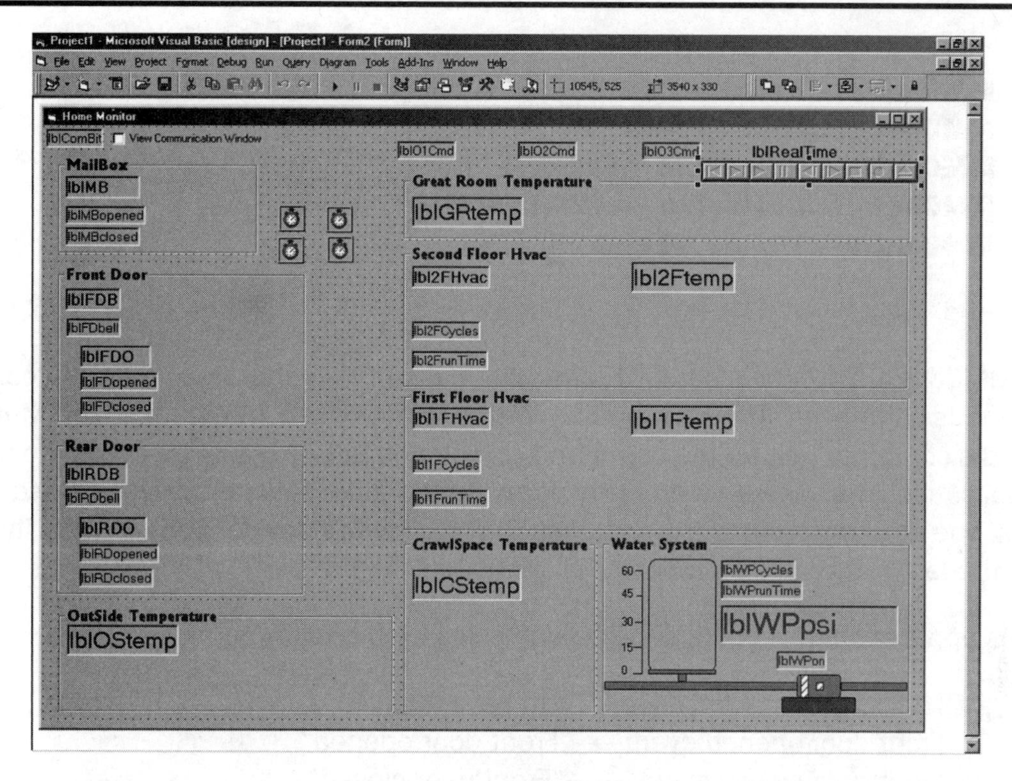

Figure 9.02

The Code

After placing the control on the form, the next step is to access the MCI device. In this case, the device type is wave audio. All new code in this chapter will be presented in **bold** type. Add the bold-type code to the Form2 load event.

```
Form2
Private Sub Form_Load()

lS1Pos = shPumpAnimate1.Left
lS2Pos = shPumpAnimate2.Left
lblComBit.Caption = " ComBit "
lblO1Cmd.Caption = " O/1 Command "
lblO2Cmd.Caption = " O/2 Command "
lblO3Cmd.Caption = " O/3 Command "
```

```
'Wave File
MMControl1.Notify = True
MMControl1.Wait = False
MMControl1.Shareable = False
MMControl1.DeviceType = "WAVEAUDIO"
MMControl1.Visible = False

End Sub
```

If MMControl1. Notify is set to true, a "Done Event" is generated when the command is completed. The MMControl1.Wait property when True causes the Multimedia control to wait for the command completion before returning control to the application. The shareable property permits other applications to access the control. The device type property specifies the type of MCI device. Control visibility is controlled by the visible property.

This application will launch specific wave files for the following door events.

1) Mailbox-open event———"You have mail"
2) Front door-opened event —"Front door opened"
3) Front door-closed event ——"Front door closed"
4) Front doorbell-pushed event-"Someone's at the front door"
5) Rear door-opened event——"Rear door opened"
6) Rear door-closed event———"Rear door closed"
7) Rear doorbell-pushed event–"Someone's at the rear door"

These files can be actual recorded wave files, wave files pulled from the Web, tone sequences, music, and so on. It's possible to make the doorbell wave file a recorded message saying that you went to the store and will return in 15 minutes, or it could be the infamous "You rang?" The associated file will be specified by the number indicated above. An ASCII "6" is the rear door closed-event wave file.

Wave files require time, and it's possible for the trigger events to occur almost simultaneously. Therefore, you need to develop a first-in/first-out mechanism to collect the wave files and issue them in the sequence that they occurred. Additionally, a flag is required to indicate whether the MMControl is free or busy generating a wave file.

Add the following variables to Module1.bas in the general-declaration section.

Module1.bas

```
Option Explicit

'Wave File
Public vWaveFileQue
Public WaveProgressFlag As Boolean


'COMMUNICATION VARIABLES AND CONSTANTS

Public Const STX As Byte = 2
Public Const ETX As Byte = 3
```

•

•

•

The variant variable vWaveFileQue will be used to collect the required wave files. This will be accomplished by concatenating the associated wave file's ASCII number to the variable when the event occurs. Another part of the program will periodically inspect this variable. If it's empty, then nothing happens; however, if it's not empty, and the MMControl is not busy, then a procedure will remove the first character from vWaveFileQue and place the remainder in vWaveFileQue. If there is no remainder, then vWaveFileQue will be set empty. The extracted character will be used to load the desired wave file into the MMControl. Subsequently, the control will be opened and instructed to play the wave file. The Boolean flag, WaveProgressFlag, is set true. Wave file completion will be indicated by the MMControl1_Done event. This event sets the WaveProgressFlag to false, which then opens the path to launch the next wave file.

Since the doorbell wave files may be somewhat lengthy, two timers are added to the form. They are named tmrFDB for the front doorbell and tmrRDB for the rear doorbell. The time interval is set to 5000, or 5 seconds for both timers. These timers permit one doorbell wave file every five seconds. The timers are enabled by the doorbell event and disabled by the time-interval event. The required code in procedure ShowFlags appears in bold type.

```
Private Sub ShowFlags()

 If ComBitflag = True Then lblComBit.BackColor = QBColor(10) Else _
lblComBit.BackColor = QBColor(14)
 If Status01flag = True Then lbl01Cmd.BackColor = QBColor(14) Else_
lbl01Cmd.BackColor = QBColor(7)
 If Status02flag = True Then lbl02Cmd.BackColor = QBColor(14) Else _
lbl02Cmd.BackColor = QBColor(7)
 If Status03flag = True Then lbl03Cmd.BackColor = QBColor(14) Else _
lbl03Cmd.BackColor = QBColor(7)


If FrontDoorBellflag = True Then
lblFDB.BackColor = QBColor(10)
lblFDB.Caption = " Front Door Bell Switch On "
lblFDbell.Caption = "Bell Rung: " & Now
If tmrFDB.Enabled = False Then
vWaveFileQue = vWaveFileQue & "4" 'Front doorbell wave
tmrFDB.Enabled = True
End If
Else: lblFDB.BackColor = QBColor(7)
lblFDB.Caption = " Front Door Bell Switch Off "
End If

If RearDoorBellflag = True Then
lblRDB.BackColor = QBColor(10)
lblRDB.Caption = " Rear Door Bell Switch On "
lblRDbell.Caption = "Bell Rung: " & Now
 If tmrRDB.Enabled = False Then
 vWaveFileQue = vWaveFileQue & "7" 'Rear doorbell wave
 tmrRDB.Enabled = True
 End If
```

•

•

•

The doorbell-timer event code simply stops the associated timer when the timer event occurs. When the timer is disabled, future doorbell wave events are possible. The code is as follows.

```
Private Sub tmrFDB_Timer()
 tmrFDB.Enabled = False
End Sub
```

```
Private Sub tmrRDB_Timer()
 tmrRDB.Enabled = False
End Sub
```

The front door, rear door, and mailbox wave file events are determined as indicated in bold in the following CheckDoorChange procedure.

```
Private Sub CheckDoorChange()

If FrontDoorOpenflag <> LastFrontDoorOpenflag Then
   LastFrontDoorOpenflag = FrontDoorOpenflag
 If FrontDoorOpenflag = True Then
 vWaveFileQue = vWaveFileQue & "2" 'Front door opened wave
     lblFDopened.Caption = " Opened: " & Now
     lblFDclosed.Caption = " Closed: ?"
 Else:
 vWaveFileQue = vWaveFileQue & "3" 'Front door closed wave
     lblFDclosed.Caption = " Closed: " & Now
 End If
End If

If RearDoorOpenflag <> LastRearDoorOpenflag Then
 LastRearDoorOpenflag = RearDoorOpenflag
 If RearDoorOpenflag = True Then
 vWaveFileQue = vWaveFileQue & "5" 'Rear door opened wave
```

```
      lblRDopened.Caption = " Opened: " & Now
      lblRDclosed.Caption = " Closed: ?"
   Else:
   vWaveFileQue = vWaveFileQue & "6" 'Rear door closed wave
      lblRDclosed.Caption = " Closed: " & Now
   End If
  End If

   If MailBoxflag <> LastMailBoxflag Then
   LastMailBoxflag = MailBoxflag
   If MailBoxflag = True Then
   vWaveFileQue = vWaveFileQue & "1" 'You have mail wave

      lblMBopened.Caption = " Opened: " & Now
      lblMBclosed.Caption = " Closed: ?"
   Else:
      lblMBclosed.Caption = " Closed: " & Now
   End If
  End If
  End Sub
```

The procedure that drives the MMControl is a new procedure called CheckWaves. The coding for this procedure is very straightforward. The Mid$ keyword is used to determine the left-most character and place it in the variable WaveNumber. Mid$ is used again to remove the left-most character from the variable vWaveFileQue. VWaveFileQue will become empty if this was the last character. The number contained in the variable WaveNumber is used to launch the appropriate wave file. Subsequently, the MMControl is opened and instructed to play. The code is as follows.

```
   Private Sub CheckWaves()

   If WaveProgressFlag = False Then
    If vWaveFileQue <> "" Then
    Dim WaveNumber
    WaveNumber = Mid$(vWaveFileQue, 1, 1)
```

```
vWaveFileQue = Mid$(vWaveFileQue, 2, Len(vWaveFileQue))

WaveProgressFlag = True

If WaveNumber = "1" Then MMControl1.FileName = "You have mail.wav"
If WaveNumber = "2" Then MMControl1.FileName = "Front door
opened.wav"
If WaveNumber = "3" Then MMControl1.FileName = "Front door
closed.wav"
If WaveNumber = "4" Then MMControl1.FileName = "You rang.wav"
If WaveNumber = "5" Then MMControl1.FileName = "Rear door
opened.wav"
If WaveNumber = "6" Then MMControl1.FileName = "Rear door
closed.wav"
If WaveNumber = "7" Then MMControl1.FileName = "You rang2.wav"

MMControl1.command = "OPEN"
MMControl1.command = "PLAY"
End If
End If
End Sub
```

The following event procedure closes an open MMControl and clears the WaveProgressFlag. Once the wave file is finished, there's no reason to leave the MMControl open.

```
Private Sub MMControl1_Done(NotifyCode As Integer)
MMControl1.command = "CLOSE"
WaveProgressFlag = False
End Sub
```

Finally, the call to the CheckWaves procedure occurs in the tmrUpdate_Timer event procedure. The code is as follows.

```
Private Sub tmrUpdate_Timer()
'Show real time
lblRealTime.Caption = Now
```

```
'At midnight send a Register clear command to the PLC
'for all elapse time timers and Cycle counters

If Hour(Now()) = "0" And Minute(Now()) = "0" And Second(Now()) < 10 Then

ClearDailyRegCmdFlag = True
Else: ClearDailyRegCmdFlag = False
End If
ShowFlags
ShowTimesAndCycles
ShowAnalog

CheckWaves 'launch wave files as required

End Sub
```

Conclusion/Exercise

This chapter demonstrates the capability of adding wave-file events to a Visual Basic project. As an exercise, you should add a wave-file event that triggers when the crawl-space temperature is less than 35 degrees Fahrenheit. The wave file will proclaim a warning of the possibility of freezing. Add a command button that acknowledges the freeze alarm and stops the wave-file output. When the temperature drops below 35 degrees, activate the alarm. Once in alarm, don't allow the alarm to clear until the temperature exceeds 38 degrees. This will prevent an oscillating alarm condition if the temperature happens to be meandering around the 35-degree Fahrenheit alarm point.

Chapter 10
DATA LOG

INTRODUCTION

This chapter looks at data logging various elements of the home-monitor program. The first part of the chapter modifies the existing program to include saving all the door events to respective files, as well as saving all analog values, cycle times, and run times to data files with every five-minute interval of time. The second part of the chapter adds the new Sub procedures required to accomplish the various data-logging tasks.

Task Description

The door-event data-log file will record each door-open or door-close event with a date/time stamp. Separate data-log files will be generated for the front door, rear door, and mailbox. Additionally, the front door and rear door data-log files will record with a date/time stamp the respective doorbell events. These data-log files will include the current month and year as part of the filename and automatically change with the progress of time. Textboxes will be added to Form2 so that the history of each door can be examined. A checkbox will added to control the visibility of all the door-event textboxes.

All the analog values, cycle numbers, and run times will be stored to a data-log file with every five-minute interval of time, based upon the PC system's real-time clock. The filename of the data file will also contain a current month and year component as part of the name.

A new timer called " tmrDataLog" will be added to the form. This timer has an interval set to the maximum of 65535, or about 65.535 seconds. This timer is used to ensure that a data log occurs only once for every five-minute interval of time.

The Code

It's time to start coding. Add three RichTextBoxes to Form2 and label them "tbMB", "tbFD", and "tbRD". Add a checkbox called "cbDoorHistory". Position these objects on the form as illustrated in Figure 10.01. The RichTextBoxes are placed over existing labels. The "cbDoorHistory" will control the visibility of the RichTextBoxes. Set the following property values for each RichTextBox: Font size 10 Bold, MultiLine = True and ScrollBars= 2-rtfVertical. Configure for vertical scroll bar only.

Note that this example is using the RichTextBox control. This control provides more advanced formatting features than the conventional TextBox control. This control isn't part of the standard toolbox tools and will need to be added with the toolbox add-component method. If desired, the standard Textbox may be used in lieu of the RichTextBox control for this application.

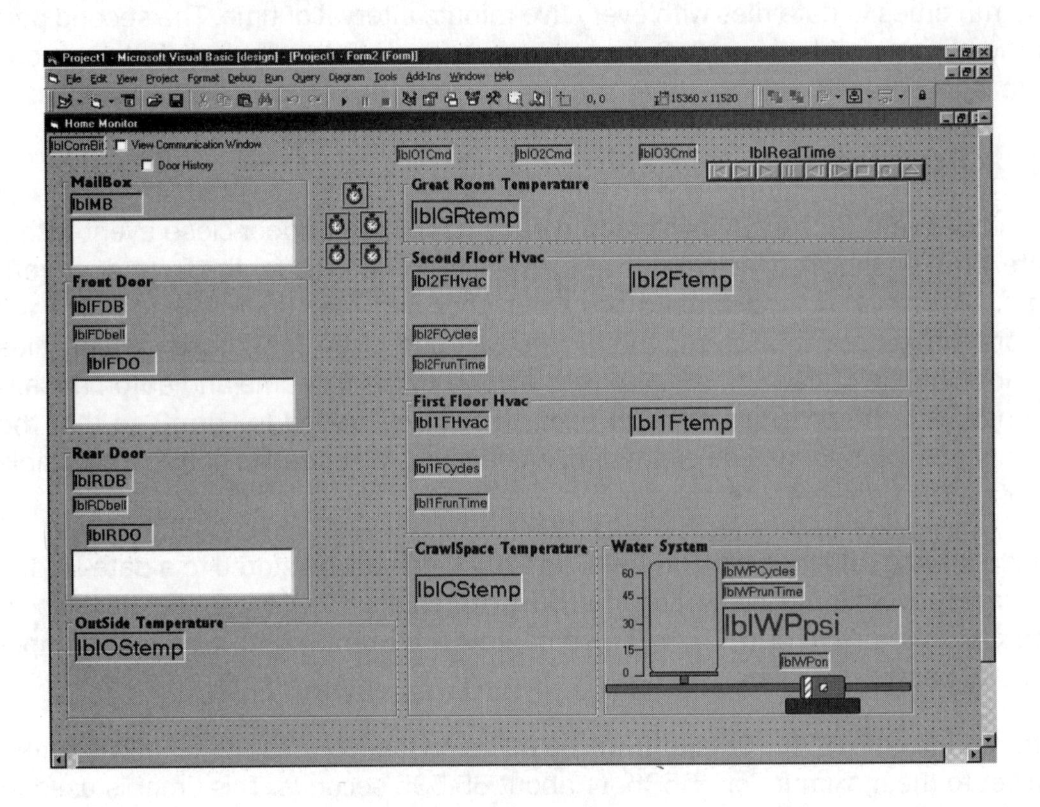

Figure 10.01

The following code shows the existing Sub procedures, with additional code shown in **bold**. All the changes occur in Form2 code.

```
Private Sub Form_Load()
  lS1Pos = shPumpAnimate1.Left
  lS2Pos = shPumpAnimate2.Left
lblComBit.Caption = " ComBit "
lblO1Cmd.Caption = " O/1 Command "
lblO2Cmd.Caption = " O/2 Command "
lblO3Cmd.Caption = " O/3 Command "
'Wave File
MMControl1.Notify = True
MMControl1.Wait = False
MMControl1.Shareable = False
MMControl1.DeviceType = "WAVEAUDIO"
MMControl1.Visible = False
cbDoorHistory.Value = 1
LoadUpTextBoxes 'with history data from files
End Sub
```

```
Private Sub tmrUpdate_Timer()
' Check for a 5 minute interval and data log
CheckDatalog
'Show real time
lblRealTime.Caption = Now
'At midnight send a Register clear command to the PLC
'for all timers and Cycle counters
If Hour(Now()) = "0" And Minute(Now()) = "0" And Second(Now()) < 10
Then
ClearDailyRegCmdFlag = True
Else: ClearDailyRegCmdFlag = False
End If
ShowFlags
ShowTimesAndCycles
ShowAnalog
CheckWaves
End Sub
```

In the Sub procedure ShowFlags, make the following changes in **bold** to the existing code.

```
If FrontDoorBellflag = True Then
lblFDB.BackColor = QBColor(10)
lblFDB.Caption = " Front Door Bell Switch On  "
lblFDbell.Caption = "Bell Rung: " & Now
 If tmrFDB.Enabled = False Then
 vWaveFileQue = vWaveFileQue & "4" 'Front Doorbell Wave
 tmrFDB.Enabled = True
 LogFrontDoorBell 'This sub procedure adds the doorbell event with
    date/time stamp
 End If
Else: lblFDB.BackColor = QBColor(7)
lblFDB.Caption = " Front Door Bell Switch Off "
End If

If RearDoorBellflag = True Then
lblRDB.BackColor = QBColor(10)
lblRDB.Caption = " Rear Door Bell Switch On  "
lblRDbell.Caption = "Bell Rung: " & Now
 If tmrRDB.Enabled = False Then
 vWaveFileQue = vWaveFileQue & "7"  'Rear Doorbell Wave
 tmrRDB.Enabled = True
 LogRearDoorBell 'This sub procedure adds the doorbell event with
    date/time stamp
 End If
Else: lblRDB.BackColor = QBColor(7)
lblRDB.Caption = " Rear Door Bell Switch Off "
End If
```

Make the following changes in **bold** to the existing Sub procedure CheckDoorChange.

```
Public Sub CheckDoorChange()
```

```
If FrontDoorOpenflag <> LastFrontDoorOpenflag Then
 LastFrontDoorOpenflag = FrontDoorOpenflag
 If FrontDoorOpenflag = True Then
 vWaveFileQue = vWaveFileQue & "2" 'Front door opened wave
  lblFDopened.Caption = " Opened: " & Now
  lblFDclosed.Caption = " Closed: ?"
 Else:
 vWaveFileQue = vWaveFileQue & "3" 'Front door closed wave
  lblFDclosed.Caption = " Closed: " & Now
 End If
 LogFrontDoor ' Time/Date Stamp the event
End If

If RearDoorOpenflag <> LastRearDoorOpenflag Then
 LastRearDoorOpenflag = RearDoorOpenflag
 If RearDoorOpenflag = True Then
 vWaveFileQue = vWaveFileQue & "5" 'Rear door opened wave
  lblRDopened.Caption = " Opened: " & Now
  lblRDclosed.Caption = " Closed: ?"
 Else:
 vWaveFileQue = vWaveFileQue & "6" 'Rear door closed wave
  lblRDclosed.Caption = " Closed: " & Now
 End If
 LogRearDoor ' Time/Date Stamp the event
End If

If MailBoxflag <> LastMailBoxflag Then
 LastMailBoxflag = MailBoxflag
 If MailBoxflag = True Then
 vWaveFileQue = vWaveFileQue & "1" 'You have mail wave
  lblMBopened.Caption = " Opened: " & Now
  lblMBclosed.Caption = " Closed: ?"
 Else:
  lblMBclosed.Caption = " Closed: " & Now
 End If
```

```
    LogMailBoxDoor ' Time/Date Stamp the event
  End If
End Sub
```

The following section of code is new code that should be appended to the existing program code. Most of the file-based Sub procedures for reading and appending data are identical. Use the copy and paste commands to cut down on the keyboard entry.

The Sub procedure DataLog converts all the analog data, cycles, and run times into strings. The strings are separated by commas and integrated into a single variant called "vData". A file called "Data*MONYR*.CVS" is opened. *MONYR* is the current month and year, such as JUN2000. The variant "vDate", along with the current date and time stamp, is appended to the file. If the file doesn't exist, a new file will be created.

```
Public Sub DataLog()
  Dim FILENUM As Byte 'temporary scratch pad variable
  Dim vData
  Dim sMOyr As String
  sMOyr = Format(Now, "MMMYYYY") 'JAN2000

'Assimilate the data
    vData = CStr(vOutsideTemp) + "," _
            + CStr(vCrawlSpaceTemp) + "," _
            + CStr(vFirstFloorTemp) + "," _
            + CStr(vSecondFloorTemp) + "," _
            + CStr(vGreatRoomTemp) + "," _
            + CStr(vWaterPressure) + "," _
            + CStr(lHvac1DailyCycle) + "," _
            + CStr(lHvac1DailyOnTime) + "," _
            + CStr(lHvac2DailyCycle) + "," _
            + CStr(lHvac2DailyOnTime) + "," _
            + CStr(lWaterPumpDailyCycle) + "," _
            + CStr(lWaterPumpDailyOnTime / 10)
```

```
On Error Resume Next
'Set file attributes to Read Only when exiting therefore
'make the attribute normal for appending data. Setting
'the attribute will stop programs like Excel from taking
'complete control of the file when it is opened for inspection
'or graphing.
SetAttr "Data" + sMOyr + ".csv", vbNormal
 FILENUM = FreeFile
 Open "Data" + sMOyr + ".csv" For Append As #FILENUM
 Print #FILENUM, Date$ + "," + Time$ + "," + vData
 Close #FILENUM
SetAttr "Data" + sMOyr + ".csv", vbReadOnly
End Sub
```

CheckDatalog checks to see if the real-time minutes end in a 0 or a 5. Every 5-minute interval of time ends in either a 0 or a 5. If it's a new 5-minute interval of time, the DataLog Sub procedure is executed, and a timer with 65.535-second time interval is enabled. This timer prevents this section of code from being executed for 65.535 seconds of time. When the timer event occurs after 65.535 seconds, the timer disables itself.

```
Private Sub CheckDatalog()
  Dim sRTmin As String
 'The Now keyword provides real time information
    sRTmin = Minute(Now)
    If tmrDataLog = False Then
      If (Right(sRTmin, 1) = "0" Or Right(sRTmin, 1) = "5") Then
      tmrDataLog.Enabled = True 'disable for 1 minute the check for 0 or 5
      DataLog
      End If
    End If
End Sub
```

```
Private Sub tmrDataLog_Timer()
    tmrDataLog.Enabled = False
End Sub
```

The following log Sub procedures save the following information in comma separated value form: date, time, and event. For the front door and back door, each door-open, door-close, and doorbell event is date/time stamped. The Mailbox only records the door-open and door-close events. The filename is constructed as (Doortype)*MONYR*.CSV. Door type is FrontDoor, RearDoor, or MailBoxDoor. *MONYR* changes with the current month and year. An example of MONYR is "May2000". In addition, each door event is appended to the respective textbox.

```
' Save the door event State
Private Sub LogFrontDoor()
  Dim FILENUM As Byte 'temporary scratch pad variable
  Dim Door As String
  Dim sMOyr As String
  Dim State As String
  sMOyr = Format(Now, "MMMYYYY") 'JAN2000
  On Error Resume Next
  'Set file attributes to Read Only when exiting therefore
  'make the attribute normal for appending data. Setting
  'the attribute will stop programs like Excel from taking
  'complete control of the file when it is opened for inspection
    'or graphing.
If FrontDoorOpenflag = True Then State = "Open" Else State = "Closed"
Door = "FrontDoor"
SetAttr Door + sMOyr + ".csv", vbNormal
  FILENUM = FreeFile
  Open Door + sMOyr + ".csv" For Append As #FILENUM
  Print #FILENUM, Date$ + "," + Time$ + "," + State
  Close #FILENUM
  'Show door event in the history text box
  tbFD.Text = tbFD.Text + Date$ + "," + Time$ + "," + State + vbLf
SetAttr Door + sMOyr + ".csv", vbReadOnly
End Sub
```

```
Private Sub LogFrontDoorBell()
  Dim FILENUM As Byte 'temporary scratch pad variable
```

```
  Dim Door As String
  Dim sMOyr As String
  sMOyr = Format(Now, "MMMYYYY") 'JAN2000
  On Error Resume Next
  'Set file attributes to Read Only when exiting therefore
  'make the attribute normal for appending data. Setting
  'the attribute will stop programs like Excel from taking
  'complete control of the file when it is opened for inspection
  'or graphing.
  Door = "FrontDoor"
SetAttr Door + sMOyr + ".csv", vbNormal
  FILENUM = FreeFile
  Open Door + sMOyr + ".csv" For Append As #FILENUM
  Print #FILENUM, Date$ + "," + Time$ + "," + "DoorBell"
  Close #FILENUM
  'Show door event in the history text box
  tbFD.Text = tbFD.Text + Date$ + "," + Time$ + "," + "DoorBell" + vbLf
 SetAttr Door + sMOyr + ".csv", vbReadOnly
End Sub
```

```
Private Sub LogRearDoorBell()
  Dim FILENUM As Byte 'temporary scratch pad variable
  Dim Door As String
  Dim sMOyr As String
  sMOyr = Format(Now, "MMMYYYY") 'JAN2000
  On Error Resume Next
  'Set file attributes to Read Only when exiting therefore
  'make the attribute normal for appending data. Setting
  'the attribute will stop programs like Excel from taking
  'complete control of the file when it is opened for inspection
  'or graphing.
  Door = "RearDoor"
 SetAttr Door + sMOyr + ".csv", vbNormal
  FILENUM = FreeFile
  Open Door + sMOyr + ".csv" For Append As #FILENUM
  Print #FILENUM, Date$ + "," + Time$ + "," + "DoorBell"
```

```
  Close #FILENUM
  'Show door event in the history text box
  tbRD.Text = tbRD.Text + Date$ + "," + Time$ + "," + "DoorBell" + vbLf
 SetAttr Door + sMOyr + ".csv", vbReadOnly
End Sub
```

```
Private Sub LogRearDoor()
  Dim FILENUM As Byte 'temporary scratch pad variable
  Dim Door As String
  Dim sMOyr As String
  Dim State As String
  sMOyr = Format(Now, "MMMYYYY") 'JAN2000
  On Error Resume Next
  'Set file attributes to Read Only when exiting therefore
  'make the attribute normal for appending data. Setting
  'the attribute will stop programs like Excel from taking
  'complete control of the file when it is opened for inspection
  'or graphing.
  If RearDoorOpenflag = True Then State = "Open" Else State = "Closed"
  Door = "RearDoor"
  SetAttr Door + sMOyr + ".csv", vbNormal
  FILENUM = FreeFile
  Open Door + sMOyr + ".csv" For Append As #FILENUM
  Print #FILENUM, Date$ + "," + Time$ + "," + State
  Close #FILENUM
  'Show door event in the history text box
  tbRD.Text = tbRD.Text + Date$ + "," + Time$ + "," + State + vbLf
  SetAttr Door + sMOyr + ".csv", vbReadOnly
End Sub
```

```
Private Sub LogMailBoxDoor()
  Dim FILENUM As Byte 'temporary scratch pad variable
  Dim Door As String
  Dim sMOyr As String
  Dim State As String
```

```
  sMOyr = Format(Now, "MMMYYYY") 'JAN2000
  On Error Resume Next
  'Set file attributes to Read Only when exiting therefore
  'make the attribute normal for appending data. Setting
  'the attribute will stop programs like Excel from taking
  'complete control of the file when it is opened for inspection
  'or graphing.
  If MailBoxflag = True Then State = "Open" Else State = "Closed"
  Door = "MailBoxDoor"
  SetAttr Door + sMOyr + ".csv", vbNormal
  FILENUM = FreeFile
  Open Door + sMOyr + ".csv" For Append As #FILENUM
  Print #FILENUM, Date$ + "," + Time$ + "," + State
  Close #FILENUM
  'Show door event in the history text box
  tbMB.Text = tbMB.Text + Date$ + "," + Time$ + "," + State + vbLf
  SetAttr Door + sMOyr + ".csv", vbReadOnly
End Sub
```

Load each door textbox with information from the current month log file. This Sub procedure is executed only during the Form_Load event.

```
Private Sub LoadUpTextBoxes()
  'Load textbox information from file
  Dim FILENUM As Byte 'temporary scratch pad variable

   Dim sMOyr As String
   Dim FileText As String
   sMOyr = Format(Now, "MMMYYYY") 'JAN2000

   'Read Mail Box History
  On Error GoTo MBBypass
    FILENUM = FreeFile
   Open "MailBoxDoor" + sMOyr + ".csv" For Input As #FILENUM

   Do While Not EOF(FILENUM)
```

```
    Line Input #FILENUM, FileText
    tbMB.Text = tbMB.Text & FileText & vbLf
    Loop
  MBBypass:
    Close #FILENUM

  CheckFrontDoor ' Load Front Door Information
  CheckRearDoor ' Load Rear Door Information

  End Sub
```

```
  Public Sub CheckFrontDoor()
    Dim FILENUM As Byte 'temporary scratch pad variable
     Dim sMOyr As String
    Dim FileText As String
    sMOyr = Format(Now, "MMMYYYY") 'JAN2000


    'Read Front Door History
   On Error GoTo FDBypass

     FILENUM = FreeFile
    Open "FrontDoor" + sMOyr + ".csv" For Input As #FILENUM

    Do While Not EOF(FILENUM)
    Line Input #FILENUM, FileText
    tbFD.Text = tbFD.Text & FileText & vbLf
    Loop
  FDBypass:
    Close #FILENUM
  End Sub
```

```
  Public Sub CheckRearDoor()
    Dim FILENUM As Byte 'temporary scratch pad variable
     Dim sMOyr As String
     Dim FileText As String
```

```
 sMOyr = Format(Now, "MMMYYYY") 'JAN2000

 'Read Rear Door History
On Error GoTo RDBypass
 FILENUM = FreeFile
 Open "RearDoor" + sMOyr + ".csv" For Input As #FILENUM

 Do While Not EOF(FILENUM)
 Line Input #FILENUM, FileText
 tbRD.Text = tbRD.Text & FileText & vbLf
 Loop
RDBypass:
 Close #FILENUM

End Sub
```

The checkbox "cbDoorHistory" controls whether the door textboxes are visible or the last door event is visible.

```
Private Sub cbDoorHistory_Click()
If cbDoorHistory.Value = 1 Then
  tbFD.Visible = True
   tbRD.Visible = True
    tbMB.Visible = True
Else
  tbFD.Visible = False
    tbRD.Visible = False
     tbMB.Visible = False
End If

End Sub
```

The final running form, with the door history checked, is provided in Figure 10.2.

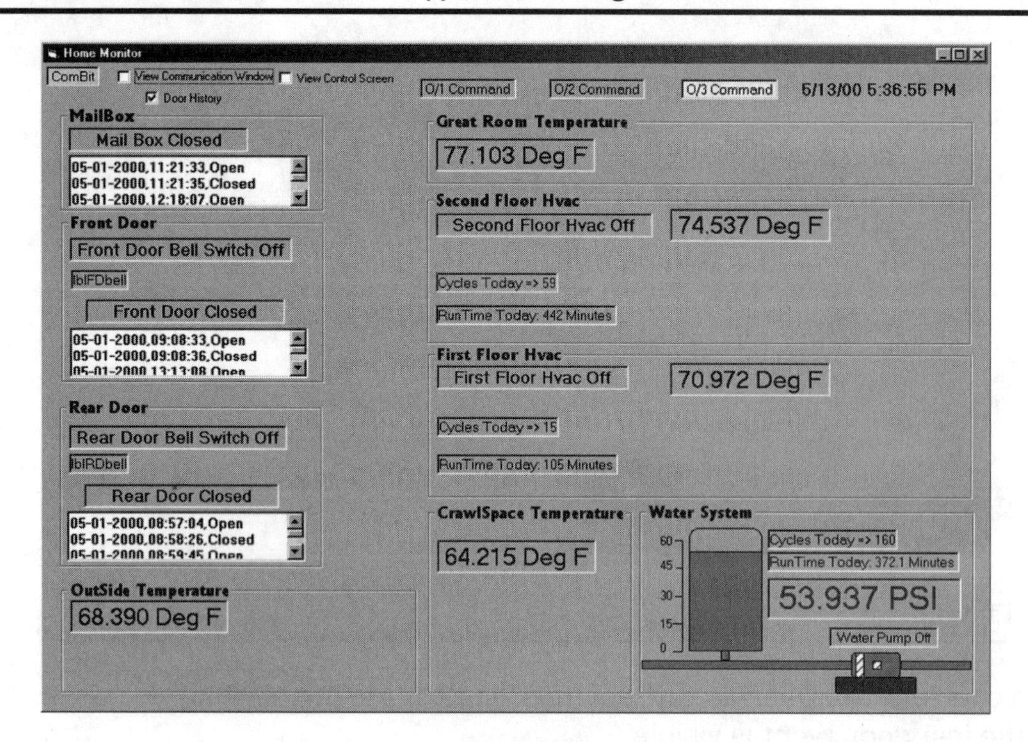

Figure 10.2

Conclusion/Exercise

The existing code loads the textboxes at the Form_Load event with the information contained in the associated file for the present month. Since the home-monitor program will be running continuously, the textboxes will become quite full as time goes by. As an exercise, modify the code to clear the textboxes at the start of a new month.

Chapter 11
GRAPHING THE DATA

INTRODUCTION

Chapter 10 created the code for generating the data-log files and provided the code to look at all the door events. These data-log files are ASCII text-based with the data separated by commas. Any text-editing program, such as Microsoft Word, WordPad, Notepad, or even the old MSDOS "Edit.com", can read these files. The CSV file extension associates these data-log files to Microsoft Excel, which will immediately separate the data in columns for analysis. Once the data is in Excel, it can be easily converted into a graphical chart. This chapter will provide the code to graphically look at all the non-door event data. In addition, the graph will be linked to the data-log file and will automatically update with the data-log file.

Task Description

Since this program will be reading the data-log files created by the home monitor, it will be a stand-alone program and not an integral part of the home-monitor program. This program will only be required to run when you want to see a graphical representation of the collected data.

This task will be accomplished by creating an ADO link to the Dataxxxxxxx.csv files. ADO is an acronym for ActiveX Data Objects. ADO was introduced by Microsoft to provide simple access to all types of data. ADO is built on an OLE DB foundation. OLE DB is the most recent Microsoft standard for connecting to both non-relational and relational databases. Microsoft ADO Data Control 6.0 will be used to provide this link.

Microsoft DataGrid Control 6.0 will be used to directly link to the ADO Data Control. The acquired data will be shown in grid form similar to a spreadsheet. This control automatically provides both horizontal and vertical scroll bars for convenient inspection of the data.

An OLE control will be placed on the form. The OLE container control allows you to add any specified object to a Visual Basic form. In this application, the object will be an embedded MSGraph object. The user-selected data contained in the DataGrid control will be sent to the MSGraph object in text form, where it will be interpreted and plotted promptly. While the program is running, MSGraph can be selected and modified at any time. Parameters such as chart type and chart size can be easily changed.

In addition, various controls will be placed on the form to provide the most convenient and effective control over the MSGraph object. So set up a new project and get started. This may sound very complicated at the moment. As you start building the code, however, a clearer picture of the process will develop.

As has been stated many times before in this book, you will be introduced to several different types of controls in this chapter. Make sure that you use Microsoft Help, as well as any other resources to supplement the learning experience. There are entire books focused just on ADO.

To the toolbar, add the Microsoft ADO Data Control 6.0 and Microsoft DataGrid components, as illustrated in Figure 11.01.

Change the caption property of Form1 to read "Home Monitor Graphical Data Window". Create a separate folder on your hard drive to store this project. Name the folder "Home Monitor Data Graphs."

The following components must be placed on Form1 for this project.

I. Six labels named and configured as follows.
   1) **Label1** not controlled change caption to "Available Data Files"
   2) **lblDataCount** default setting just change font to bold
   3) **lblDataSize** AutoSize=True; BorderStyle="1-fixed Single"
      ToolTipText ="DOUBLE CLICK ON TO CHANGE"
   4) **lblDataSource** default setting
   5) **lblEndDataPoint** AutoSize=True; BorderStyle="1-fixed Single"
   6) **lblRealTime** AutoSize=True; BorderStyle="1-fixed Single"

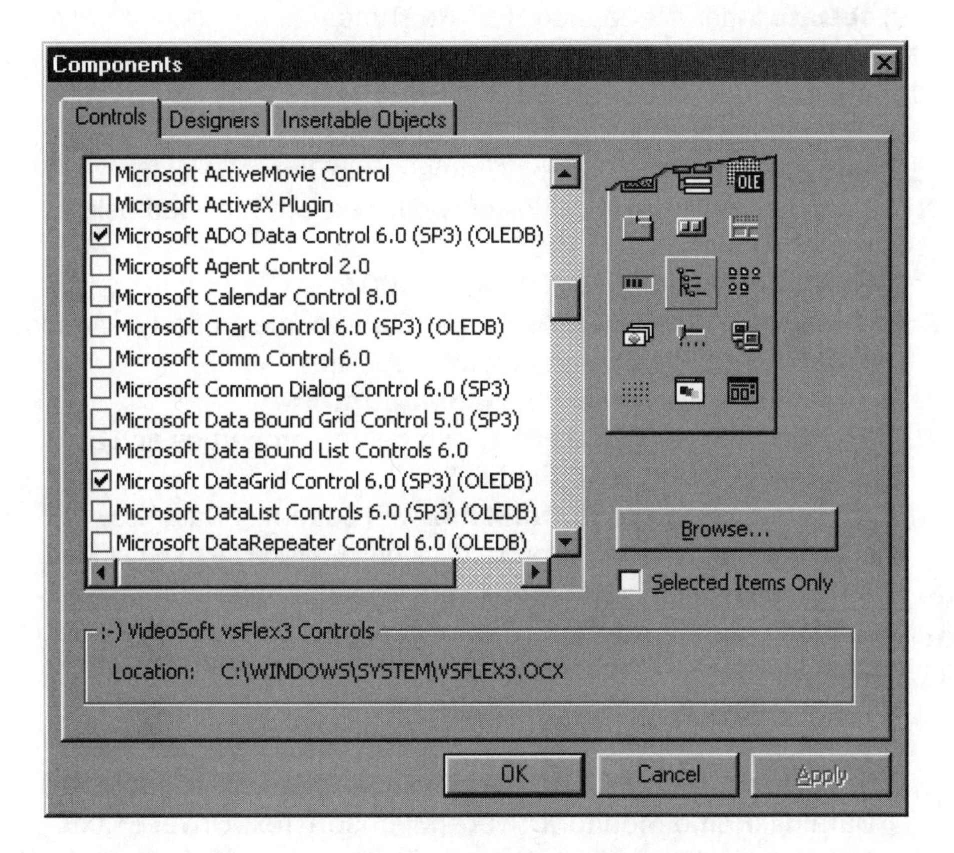

Figure 11.01

II. Thirteen checkboxes with the following names and captions. Note that cbData is a checkbox array and that an index number is provided for each data type.

1) **cbAuto5MinUpdate** caption “Auto 5 Minute Update”
2) **cbData** index #1 Caption “Outside Temp.”
3) **cbData** index #2 Caption “CrawlSpaceTemp.”
4) **cbData** index #3 Caption “Floor1 Temp.”
5) **cbData** index #4 Caption “Floor2 Temp.”
6) **cbData** index #5 Caption “GreatRoom Temp.”
7) **cbData** index #6 Caption “Water Pressure”
8) **cbData** index #7 Caption “F1 Cycles”

9) **cbData** index #8 Caption "F1 RunTime"
10) **cbData** index #9 Caption "F2 Cycles"
11) **cbData** index #10 Caption "F2 RunTime"
12) **cbData** index #11 Caption "Pump Cycles"
13) **cbData** index #12 Caption "Pump RunTime"

III. The names and interval values of two timers are provided below.
1) **tmr100ms** interval = 100
2) **tmrCheckDataSource** interval = 400

IV. One DataGrid named **DataGrid1** with DataSource property = Adodc1

V. One OLE named **OLE1** with default properties

VI. One command Button named **cmdManUpDate**

VII. One textbox named **tbDataSize** with default properties and a ToolTipText of "Must be a Number"

VIII. One FileListBox named **flbDataFiles** with pattern of Data*.csv. Change the ToolTipText to "Double Click on Desired File to be plotted."

IX. One HScrollBar named **HScroll1** with default properties

X. One Adodc named **Adodc1**. The following property changes are required to Adodc1.
1) CommandType = 2-adCmdTable
2) Provider=MSDASQL.1;Persist Security Info=False;Extended Properties="DefaultDir=C:\Book\Book Programs\Chapter 10 Datalog\Home Monitor;Driver={Microsoft Text Driver (*.txt; *.csv)};DriverId=27;Extensions=None,asc,csv,tab,txt;FIL=text;FILEDSN=C:\Book\Book Programs\Chapter 10 Datalog\Home Monitor\HomeMonitor.dsn;MaxBufferSize=2048;MaxScanRows=8;PageTimeout=5;SafeTransactions=0; Threads=3;UID=admin;UserCommitSync=Yes;"
3) RecordSoure = DataApr2000.csv

The best way to create the connection string is to use the Adodc1 intrinsic build function. Activate the custom property in the Adodc1 property window. You will see the dialog box shown in Figure 11.02. Clicking on the Build Command button leads to another dialog box, as illustrated in Figure 11.03.

Subsequently pressing the Build Command button on the second dialog window leads to the Select Data Source dialog box shown in Figure 11.04. You now have a choice to browse for an existing Data Source or to create a new Data Source. If

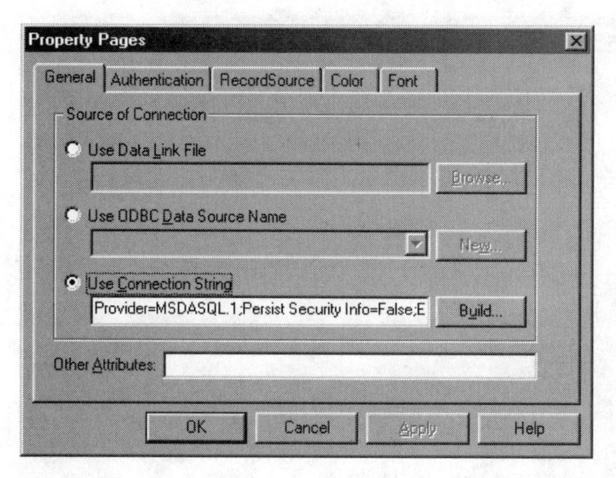

Figure 11.02

you're working with the files supplied with this book's CD-ROM, then use the folder-up icon to browse for the CSV-based datalog file created in the "Chapter 10 DataLog/Home Monitor/" folder. As the home-monitor program evolves through later chapters of this book, you will need to change the DSN path to the CSV file of each respective new chapter. The most current data-logging file will occur in the folder of the chapter in which you're working.

Ideally, all the data-log files should be stored in a dedicated folder contained in the root directory of the hard drive. To create a New DSN name, click on the New Command button found in Figure 11.04. This action opens the door to create just about any desired data source. Figure 11.05 illustrates several of the available drivers. You can use any desired name for the DSN file. In this case, the chosen name is "HomeMonitor.DSN".

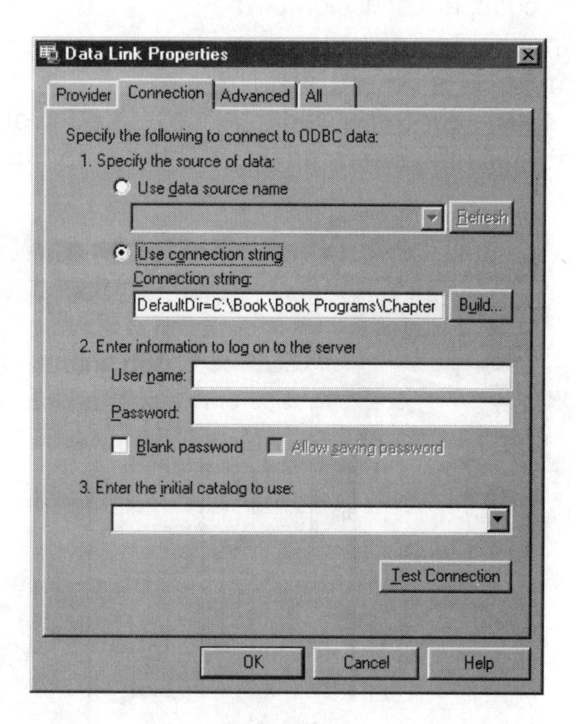

Figure 11.03

If you open the newly created DSN file with WordPad, you will see the textual information presented in Figure 11.06. The acronym ODBC stands for Open DataBase Connectivity, which is a standard programming language interface used to connect to a variety of data sources.

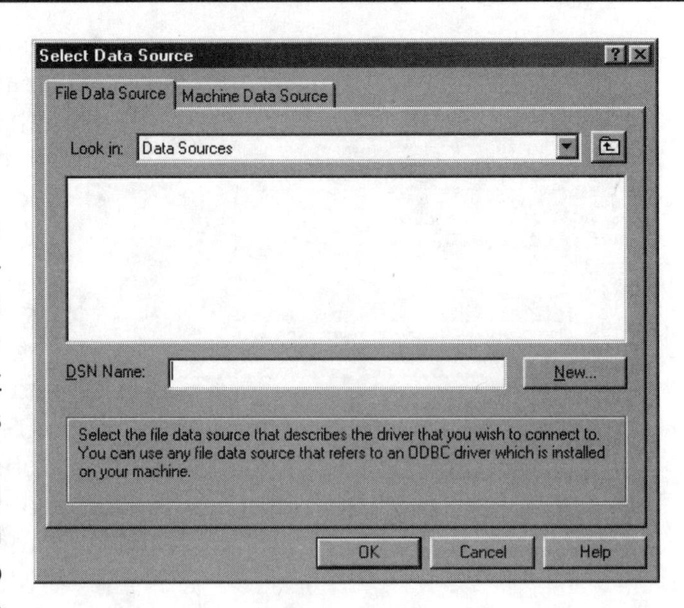

Figure 11.04

It's time to get back on the project track. Arrange the previously mentioned required controls on Form, as shown in Figure 11.07. Timer placement is not critical because timers are not visible at run time. Take note that the label "lblDataSource" is placed in OLE1. Before you move on to the code, a discussion of the functionality of the various controls is in order. The DataGrid control is tied to ADO Data Control, which in turn is linked to the active data-log file generated by the home-monitor program developed in Chapter 10.

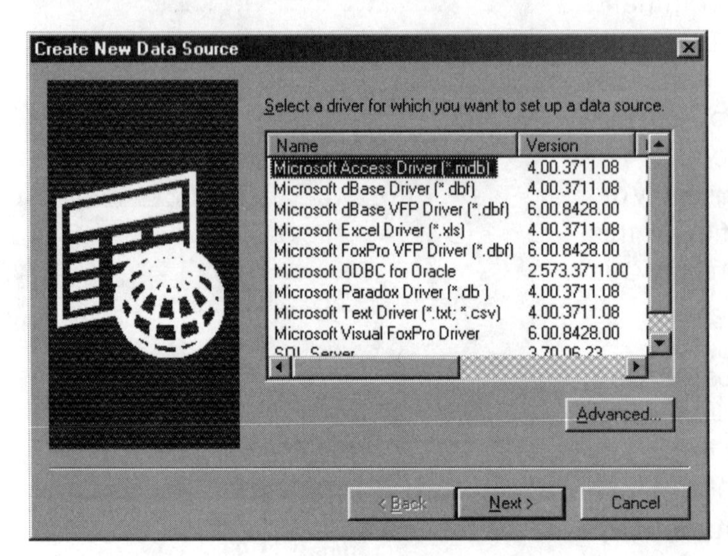

Figure 11.05

The home-monitor program records 12 items. Checkboxes are provided in order to allow the specified item or items to be included in the graph. One selection can be graphed, any combination of selections can be graphed, or they can all be graphed.

The home-monitor program records the collected data every five minutes. The worst-case

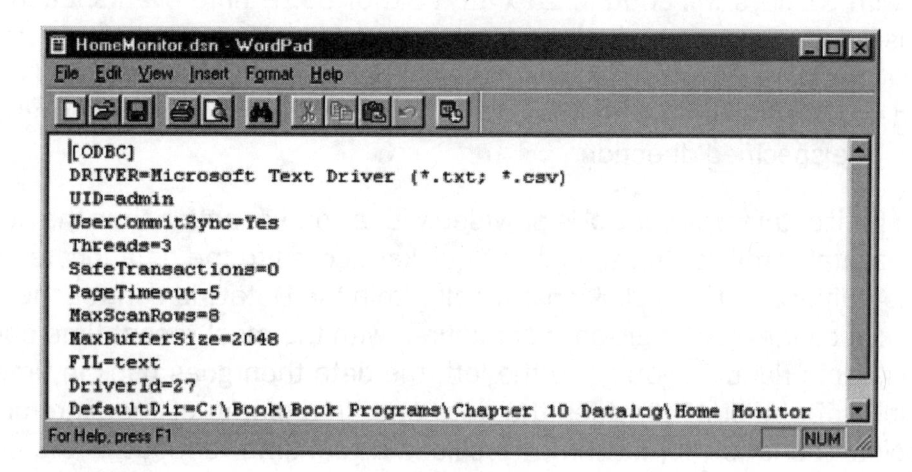

Figure 11.06

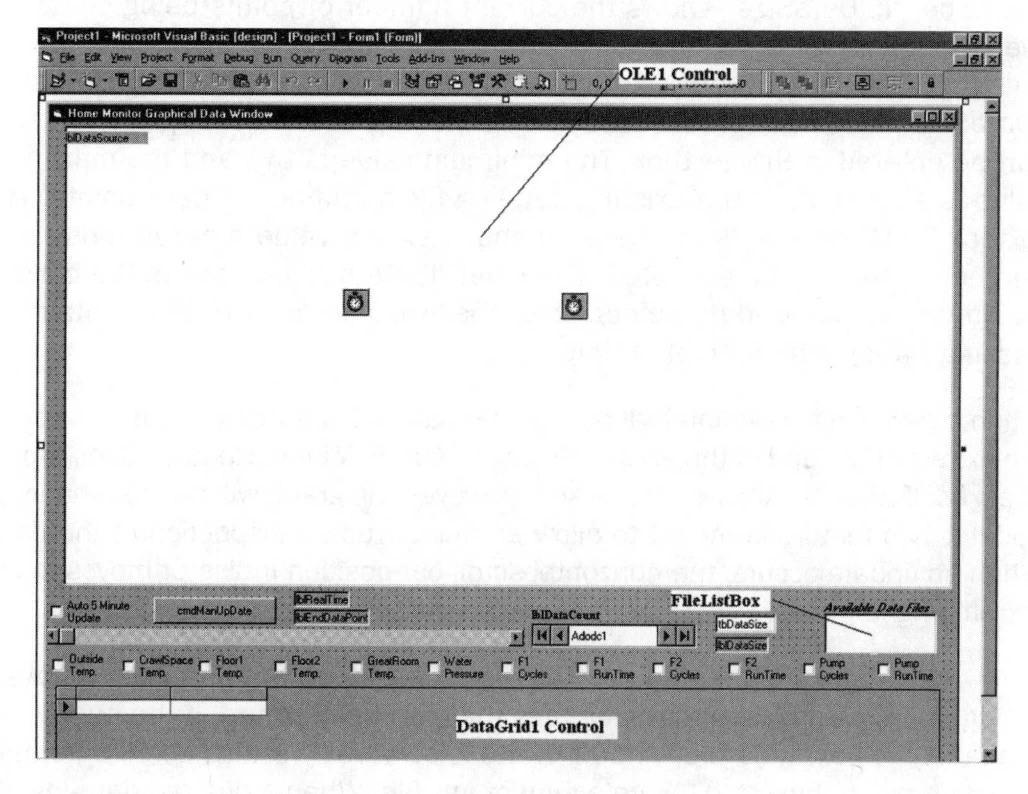

Figure 11.07

month with 31 days will contain 24 x 12 x 31, or 8928 time events. Each event contains a date/time stamp and the 12 recorded items. The extreme end horizontal scroll members of the Adodc1 will navigate between the beginning and the end of the data. The inside horizontal scroll buttons of Adodc1 provide a single-step action in the specified direction.

Another horizontal scroll control is provided with a coarser action than the Adodc1. The horizontal scroll control will permit quicker access to the data located in the middle. Additionally, if you click on any cell within the DataGrid control, the Graph will track accordingly. The graph is configured with the most recent time oriented to the extreme right. As you go to the left, the data then goes back in time. The label "lblEndDataPoint" provides the date and time stamp of the extreme right data point. The label "lblRealTime" provides a real-time readout of the current date and time.

The Label "lblDataSize" shows the current number of points being plotted in the graph. This value has a default setting of 20 data points established during the Form_Load event. If this label is double-clicked on, the textbox "DataSize" becomes both focused and visible. Any desired change in data size can be entered in this textbox. The minimum value is two and the maximum value is a function of the existing data-log file's number of data points. The textbox ToolTipText is "Must be a number", so the value entered must be a number in order to be accepted. The label "lblDataCount" shows the current maximum number of data points. This label will increment in value with each passing five-minute interval of time.

When a new five-minute interval occurs, the data will automatically be plotted at the extreme right end of the graph—that is, if "Auto 5 Minute Update" CheckBox is checked. If this checkbox is unchecked, however, the graph will not automatically update. This feature is meant to allow an uninterrupted inspection of the data. When an update occurs, the horizontal scroll bar-position indicator moves to the extreme right.

A new data-log file is created monthly. The FileListBox automatically shows a listing of all the available data-log files that can be graphed. If the number of available files exceeds the window of the FileListBox, vertical scroll automatically appears to facilitate the selection of any file. When a desired data-log file is double-clicked on, it becomes the DataSource for the Adodc control and will then be plotted. Both the label caption for the Label "lblDataSource" and Adodc1 caption will indicate the filename of the current data source being plotted.

Considering the graphing power this form contains, you will be surprised at the small amount of code required for the task. Essentially, the controls do most of the work. The ADO object dynamically links to the data file. The DataGrid control is a data-aware control that automatically contains the data linked by the ADO control.

The DataGrid has a property called BookMark. This property can either set or return the active row in a DataGrid control. When BookMark is set to a valid value in code, the row associated with that value becomes the current row. The grid automatically adjusts its display to bring the new current row into view. The ApproxCount property provides a value representing the maximum number of active rows in a DataGrid control.

The DataGrid properties BookMark and ApproxCount are used in conjunction with the DataGrid celltext method to retrieve the data from a selected cell. For-Next loops are used to assemble all the specified data into one sizable text string. This text string is then routed through the OLE control to MsGraph, and the OLE subsequently instructs Msgraph to plot the data. This text string is then collected by an associated MsGraph datasheet. Tab Commands are sent instructing the datasheet to jump to a new cell. A linefeed is issued to start a new line. After the data is successfully entered, the data is plotted.

The Code

The following code is straightforward. You shouldn't have any great difficulty working through it. Use Microsoft Help for any items that you don't understand. Set breakpoints where necessary and single-step through the code to gain clarification.

```
Option Explicit


Public iNumberOfPlotPoints As Integer
```

```
Sub Form_Load()

  Dim sMOyr As String
  Dim sDataSource As String
  sMOyr = Format(Now, "MMMYYYY") 'JAN2000
  sDataSource = "Data" + sMOyr + ".csv"

 With Adodc1
    .CommandType = adCmdTable
     .RecordSource = sDataSource
     .Refresh 'get data
 End With

 DataGrid1.ColumnHeaders = True
 Set DataGrid1.DataSource = Adodc1
 ColumnCaption ' load data identifying headers

  ' load defaults and initialize the controls
  tbDataSize.Visible = False
  OLE1.Format = "CF_TEXT"   ' Set the file format to text.
  OLE1.SizeMode = 2 ' 2=Autosize. 1=STRETCH
  OLE1.CreateEmbed "", "msGRAPH"
  DataGrid1.Bookmark = DataGrid1.ApproxCount
  HScroll1.Value = 32767
  iNumberOfPlotPoints = 20
  lblDataSize.Caption = " Plot Data Size = " + CStr(iNumberOfPlotPoints)
  cbData(1).Value = 1
  cbAuto5MinUpdate.Value = 1
  ' set file path for File List Box Control
  flbDataFiles.Path="C:\Book\Book Programs\Chapter 10 Datalog\Home
  Monitor"
   cmdManUpDate.Caption = "Manual Update"


 End Sub
```

```
Private Sub ColumnCaption()

 DataGrid1.Columns(0).Caption = "Date"
 DataGrid1.Columns(1).Caption = "Time"
 DataGrid1.Columns(2).Caption = "Outside Temp"
 DataGrid1.Columns(3).Caption = "Crawlspace Temp"
 DataGrid1.Columns(4).Caption = "Floor1 Temp"
 DataGrid1.Columns(5).Caption = "Floor2 Temp"
 DataGrid1.Columns(6).Caption = "Great Room Temp"
 DataGrid1.Columns(7).Caption = "Water Pressure"
 DataGrid1.Columns(8).Caption = "F1 Cycles"
 DataGrid1.Columns(9).Caption = "F1 Run Time"
 DataGrid1.Columns(10).Caption = "F2 Cycles"
 DataGrid1.Columns(11).Caption = "F2 Run Time"
 DataGrid1.Columns(12).Caption = "Pump Cycles"
 DataGrid1.Columns(13).Caption = "Pump Run Time"

End Sub
```

```
Private Sub cbData_Click(Index As Integer)
' Any time a data checkbox is clicked replot the data
GOGRAPH
End Sub
```

```
Private Sub GOGRAPH()
On Error Resume Next

Dim vMsGraph  ' Declare variables.
Dim sNewLine As String * 1
Dim sTab As String * 1
Dim Row As Integer
Dim vChartArray(12)
Dim i As Integer
  sTab = Chr(9)   ' Tab character.
```

```
sNewLine = Chr(10)   ' Newline character.
' Create data to replace default Graph data.

If DataGrid1.Bookmark - iNumberOfPlotPoints <= 0 Then
DataGrid1.Bookmark = DataGrid1.Bookmark + iNumberOfPlotPoints
End If
```

'Concatenate the date/time data of the specified plot data into a text String and store in 'vChartArray(0).

'Assemble the data so that the older time based date is the first in line. This creates a 'normal progressive time sweep from left to right. Older time on the left.

```
 For Row = iNumberOfPlotPoints To 0 Step -1
 vChartArray(0)=vChartArray(0)+sTab+
DataGrid1.Columns(1).CellText(DataGrid1.Bookmark - Row) _
        + "," + DataGrid1.Columns(0).CellText(DataGrid1.Bookmark - Row)
 Next Row
 vChartArray(0) = vChartArray(0) + sNewLine
```

'Gather the data and place in an array based upon a check in the checkbox and the 'specified data plot size window.

```
 For i = 1 To 12
  If cbData(i).Value = 1 Then 'skip if the associated checkbox is unchecked
   For Row = iNumberOfPlotPoints To 0 Step -1
   vChartArray(i) = vChartArray(i) + sTab _
     + DataGrid1.Columns(i + 1).CellText(DataGrid1.Bookmark - Row)
   Next Row
   vChartArray(i) = vChartArray(i) + sNewLine
  End If
 Next i

vMsGraph = ""
For i = 0 To 12
```

```
vMsGraph = vMsGraph + vChartArray(i)
Next i

 ' VbOLEHide -3 For embedded objects, hides the application that cre-
ated the object.

 ' Send the data using the DataText property.
 ' Activate MSGRAPH as hidden.
OLE1.DoVerb -3
  If OLE1.AppIsRunning Then
    OLE1.DataText = vMsGraph
' Update the object.
OLE1.Update
  Else
    MsgBox "Graph isn't active."
  End If
End Sub
```

```
Private Sub lblDataSize_DblClick()
'show text box for new entry: load with current number of plot points
With tbDataSize
 .Visible = True
 .SetFocus
 .Text = iNumberOfPlotPoints
 End With

End Sub
```

```
Private Sub tbDataSize_KeyPress(KeyAscii As Integer)
'check for a carriage return which indicates a new entry
If KeyAscii = Asc(vbCr) Then
 If IsNumeric(tbDataSize.Text) = True Then 'quick check for a numeric value
  iNumberOfPlotPoints = CInt(tbDataSize.Text)
  ' Set minimum to 2 and maximum to current number of data points
  If iNumberOfPlotPoints = 0 Then iNumberOfPlotPoints = 2
```

```
    If iNumberOfPlotPoints > DataGrid1.ApproxCount Then
iNumberOfPlotPoints = DataGrid1.ApproxCount
   lblDataSize.Caption = "Plot Data Size = " + CStr(iNumberOfPlotPoints)
  tbDataSize.Visible = False
  GOGRAPH
 End If
End If
End Sub
```

```
Private Sub flbDataFiles_DblClick()

'change the ADO record source files
If flbDataFiles.FileName <> "" Then
 With Adodc1
    .RecordSource = flbDataFiles.FileName
    .Refresh
 End With
End If

End Sub
```

```
Private Sub HScroll1_Change()
If HScroll1.Value < 1000 Then HScroll1.Value = 1000
DataGrid1.Bookmark=(HScroll1.Value / 32767) * DataGrid1.ApproxCount
'detect minimum and maximum
If DataGrid1.Bookmark < iNumberOfPlotPoints And DataGrid1.ApproxCount>
 iNumberOfPlotPoints Then
DataGrid1.Bookmark = (DataGrid1.ApproxCount* 0)+ iNumberOfPlotPoints
  End If

IfDataGrid1.Bookmark>DataGrid1.ApproxCountThen
DataGrid1.Bookmark=DataGrid1.ApproxCount
```

```
End Sub
```

```
Private Sub tmr100ms_Timer()
'this timer checks for changes to the Horizontal Scroll bar and only performs
'10 possible graph changes per second
Static lLastDataGrid1Bookmark As Long
If lLastDataGrid1Bookmark <> DataGrid1.Bookmark Then
lLastDataGrid1Bookmark = DataGrid1.Bookmark
GOGRAPH
lblEndDataPoint.Caption = "Last Data Point Date/Time: " &
DataGrid1.Columns(0).CellText(DataGrid1.Bookmark) + " " +
DataGrid1.Columns(1).CellText(DataGrid1.Bookmark)
'Show Scroll bar proportionally to the available data
HScroll1.Value = (DataGrid1.Bookmark / DataGrid1.ApproxCount) *
32767
End If

lblRealTime.Caption = Now
lblDataSource.Caption = Adodc1.RecordSource
lblDataCount.Caption = "Total Data Points: " & DataGrid1.ApproxCount
Adodc1.Caption = Adodc1.RecordSource

End Sub
```

```
Private Sub cmdManUpDate_Click()
' command Button update
Adodc1.Refresh
ColumnCaption
DataGrid1.Bookmark = DataGrid1.ApproxCount

End Sub
```

```
Private Sub tmrCheckDataSource_Timer()
'Check for a change in the Datasource which indicate new data
```

```
Static AutoRefreshFLAG As Boolean
Dim A As Integer
Dim sMOyr As String
Static sLastMOyr As String
Dim sRTmin As String
 'The Now keyword provides real time information
 sRTmin = Minute(Now)

'if the auto update is enabled start looking for new data 3 seconds after any
'new five minute interval

If cbAuto5MinUpdate.Value = 1 Then
   If (Right(sRTmin, 1) = "0" Or Right(sRTmin, 1) = "5") Then
   A = Val(Format(Time$, "Ss"))
   If A < 3 Then AutoRefreshFLAG = False

    If A > 3 And AutoRefreshFLAG = False Then
    AutoRefreshFLAG = True
     Adodc1.Refresh
      ColumnCaption
      HScroll1.Value = 32767
       DataGrid1.Bookmark = DataGrid1.ApproxCount
'check to see if any new data files exists—File List Box
        flbDataFiles.Refresh
    End If
  End If
End If

'Check for a new month roll over and change record source
'wait at least 30 minutes

A = Val(Format(Time$, "n")) ' with the format command "n" displays minutes
  sMOyr = Format(Now, "MMMYYYY") 'JAN2000
 If sLastMOyr = "" Then sLastMOyr = sMOyr
 If sLastMOyr <> sMOyr And A > 33 Then
```

```
sLastMOyr = sMOyr
 With Adodc1
     .RecordSource = "Data" + sMOyr + ".csv"
  .Refresh
 End With
End If

End Sub
```

Sample Graph Screens

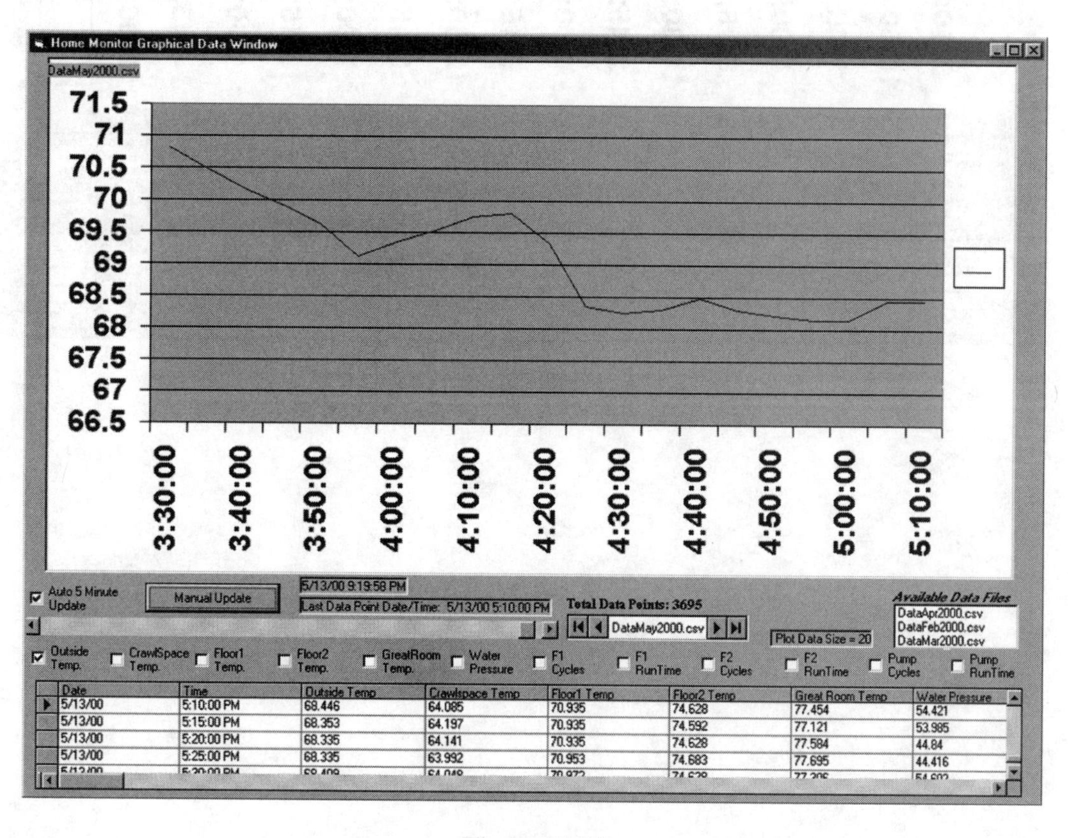

Date	Time	Outside Temp	Crawlspace Temp	Floor1 Temp	Floor2 Temp	Great Room Temp	Water Pressure
5/13/00	5:10:00 PM	68.446	64.085	70.935	74.628	77.454	54.421
5/13/00	5:15:00 PM	68.353	64.197	70.935	74.592	77.121	53.985
5/13/00	5:20:00 PM	68.335	64.141	70.935	74.628	77.584	44.84
5/13/00	5:25:00 PM	68.335	63.992	70.953	74.683	77.695	44.416

Figure 11.08
A Single Variable with a Plot Size of 20 Points

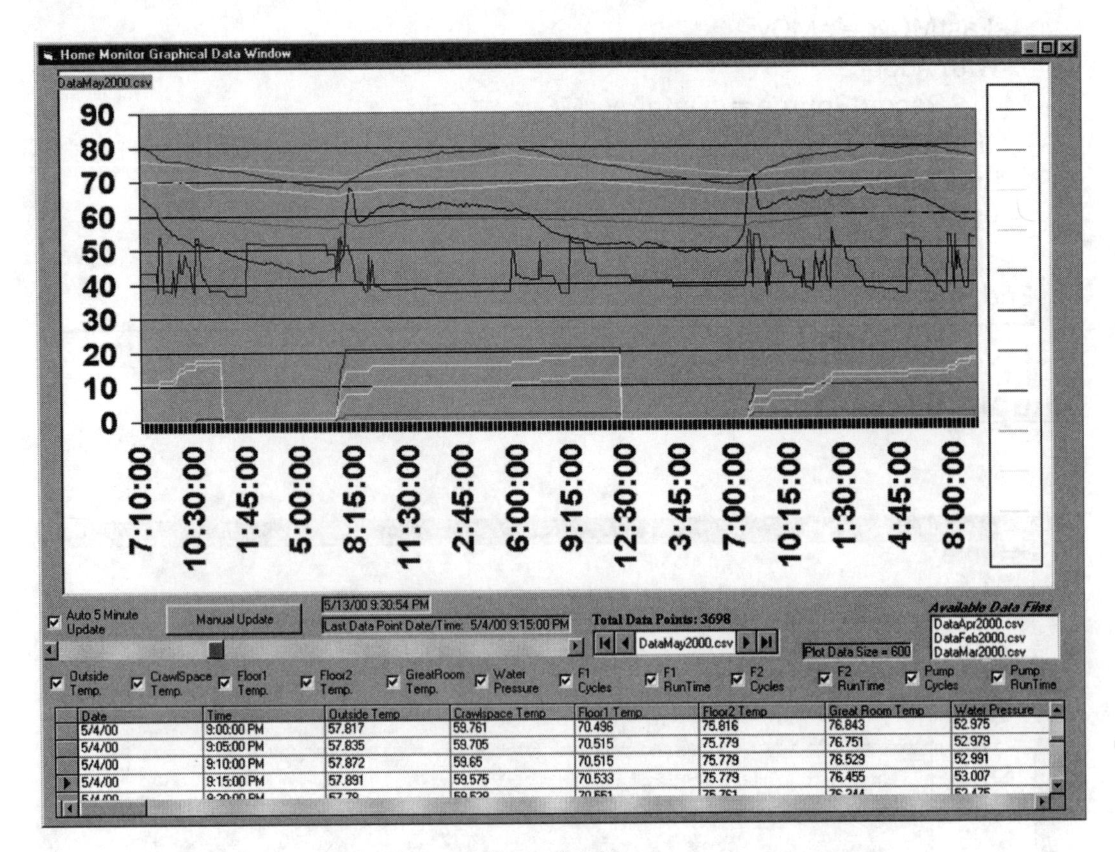

Figure 11.09
All 12 Variables with a Plot Data Size of 600 Records

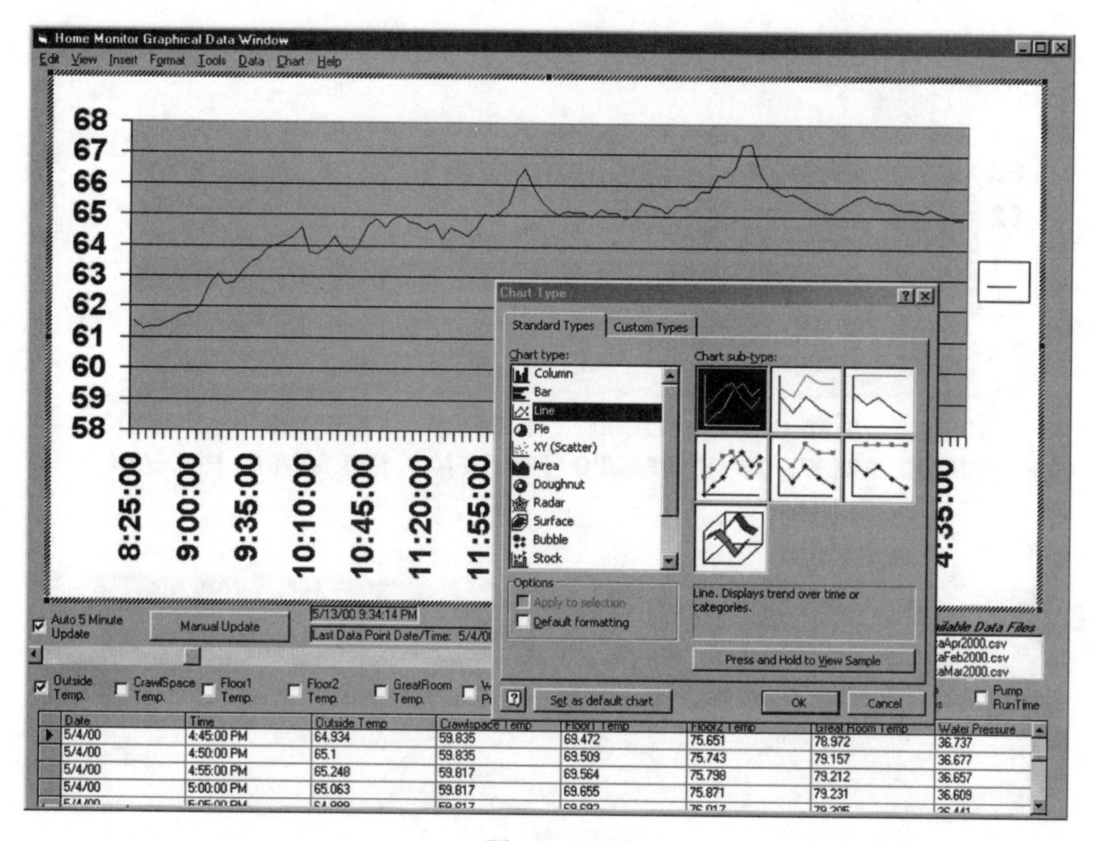

Figure 11.10
The Many Chart Options Available with an OLE Link to MSGraph

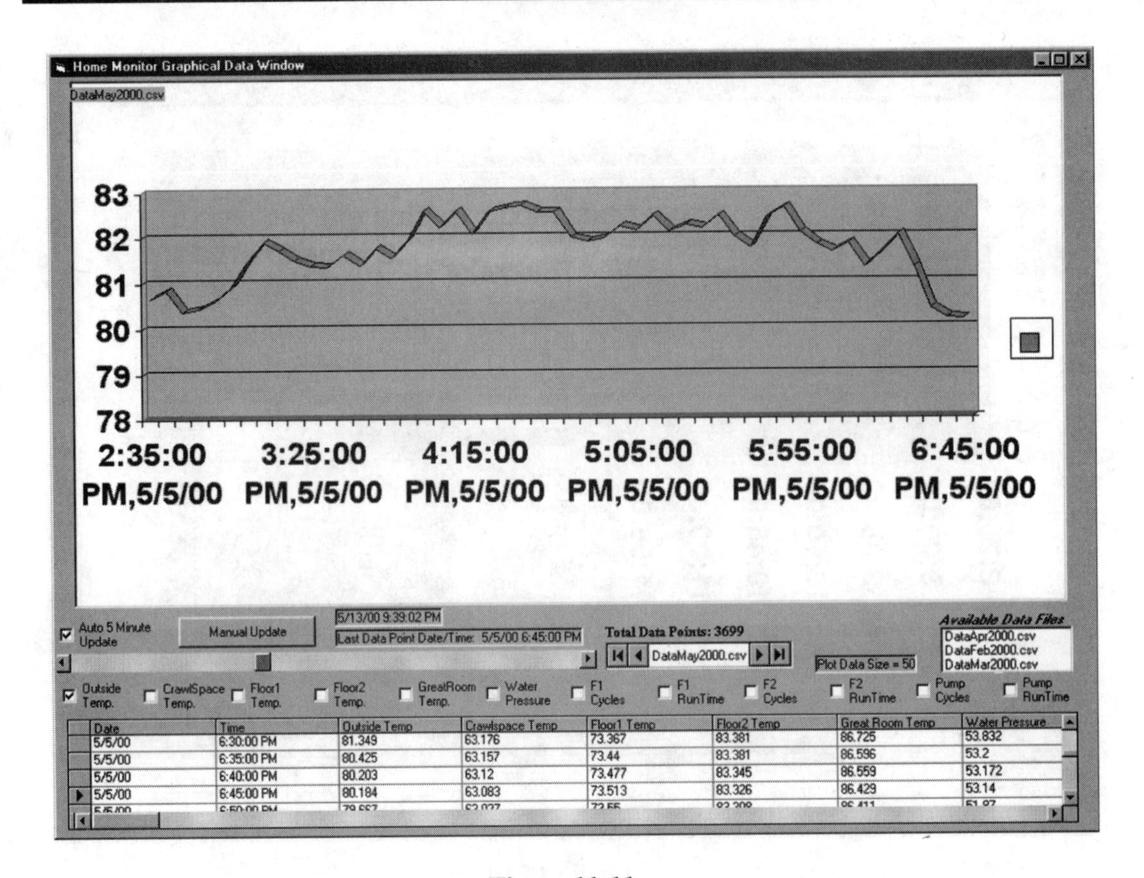

Figure 11.11
A Chart Type of a Three-Dimensional Line Trend

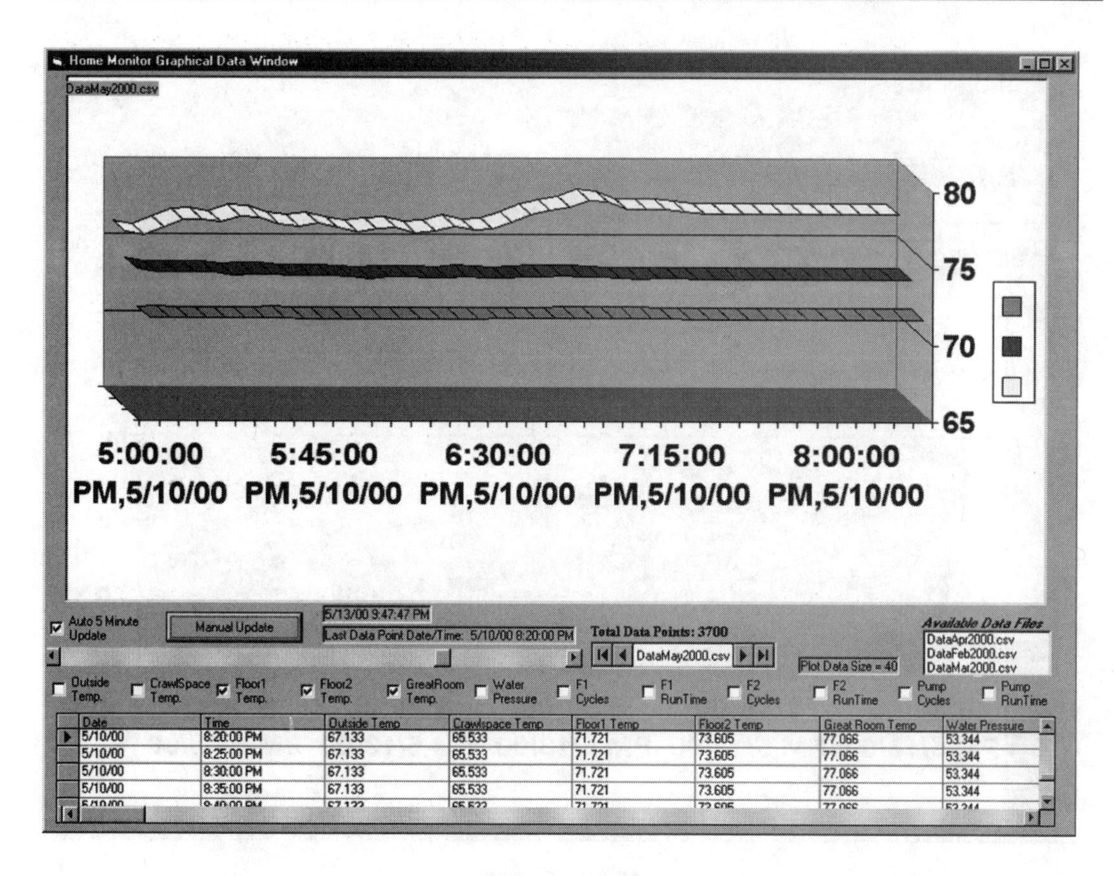

Date	Time	Outside Temp	Crawlspace Temp	Floor1 Temp	Floor2 Temp	Great Room Temp	Water Pressure
5/10/00	8:20:00 PM	67.133	65.533	71.721	73.605	77.066	53.344
5/10/00	8:25:00 PM	67.133	65.533	71.721	73.605	77.066	53.344
5/10/00	8:30:00 PM	67.133	65.533	71.721	73.605	77.066	53.344
5/10/00	8:35:00 PM	67.133	65.533	71.721	73.605	77.066	53.344

Figure 11.12
A Graph of Three Variables as a Three-Dimensional Line Trend

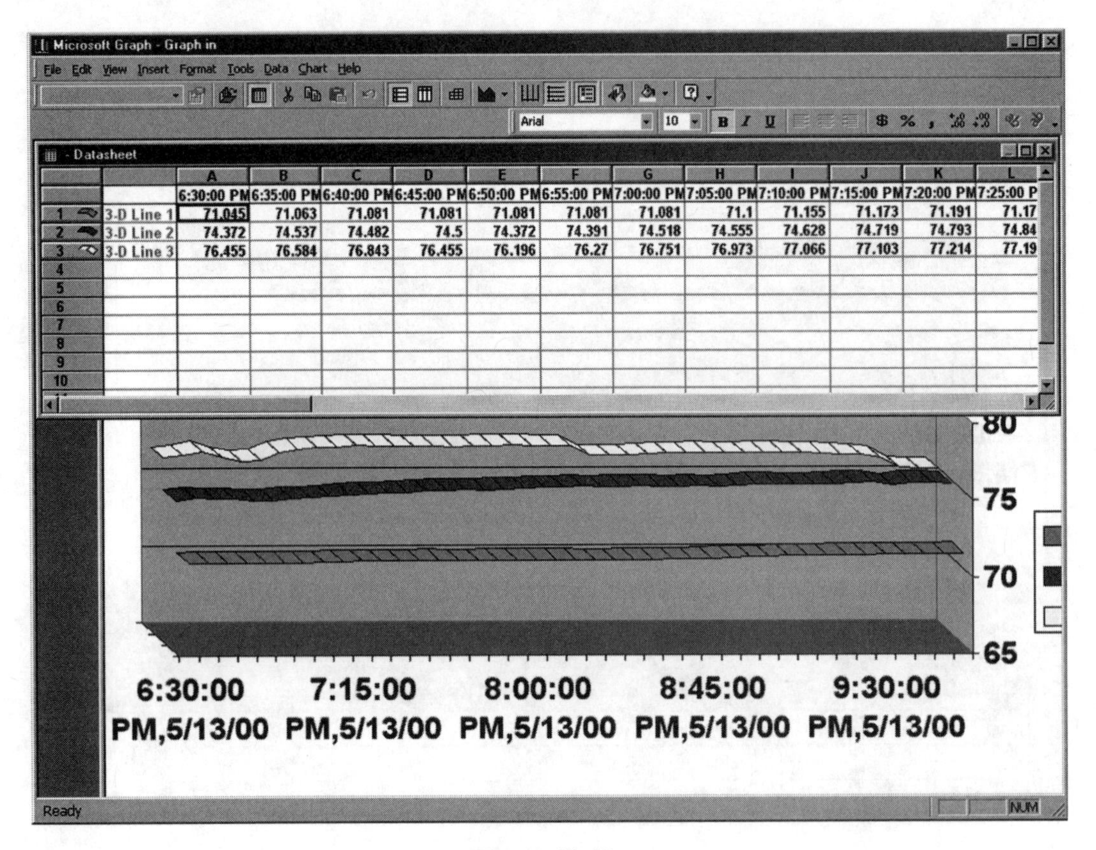

Figure 11.13
A View of the MSGraph Datasheet

Conclusion/Exercise

As you can see from these sample views, this is a very powerful data-graphing package that has been constructed with relative ease using Visual Basic. As an exercise, modify the program so the column-heading captions are routed to Msgraph in order to provide a labeled color-coded legend.

Chapter 12
SERIAL ALPHANUMERIC DISPLAY

INTRODUCTION

This chapter discusses interfacing to a serial alphanumeric display. An alphanumeric display will be used to remotely display the data collected by the home-monitor program in textual form. The information will be transmitted serially in ASCII.

Task Description

Serial alphanumeric displays are available in a number of different types, different interface configurations, and different display configurations. The most popular types are LED (light emitting diode), LCD (liquid crystal display), Plasma, and Vacuum Fluorescent (VF). Both serial and parallel interface configurations are available. Single-line and multiple-line display models are available with an assortment of number of characters per line. The serial-based modules have a microprocessor on board that drives the display and services the communications. You simply send a serial text-message string and the message is automatically displayed. Control codes are available that can change the intensity of the display and serve other miscellaneous functions.

For this project, a Vacuum Fluorescent display with four lines (each line contains 20 characters) was selected. The display is an Industrial Electronic Engineers, Inc. Flip 4-Line x 20-Character 5 x 7 Dot Matrix Vacuum Fluorescent Display, with an RS-232-C serial interface. Although this model supports both write and read requests, this application will only write to the display module. The manufacturer of the display was kind enough to supply a photograph for publication in this book, which shows a series of Vacuum Fluorescent display modules. Industrial Electronic Engineers, Inc. in no way implies or states any certification or approval of the material discussed in this book. Figure 12.01 shows a number of different types of Vacuum Fluorescent displays on the market today.

Figure 12.01

Interconnection to this device is minimal. Since this project is using the Vacuum Fluorescent display for display purposes only and not for data entry, two wires are required to provide the one-way communication link. A five-volt power supply is required to power the device. Multiple displays can be driven by simply paralleling the two communication wires. Figure 12.02 shows a multiple display interconnection.

This project display has 80 characters grouped into four lines, with 20 characters per line. The display has a marker called a cursor. When an ASCII text string is sent to the display, the first character is placed at the cursor position. The cursor subsequently shifts to the right, placing the received characters as it processes the message. The cursor automatically drops down to the start of a new line once the last character of the current line is placed. Special control codes are available,

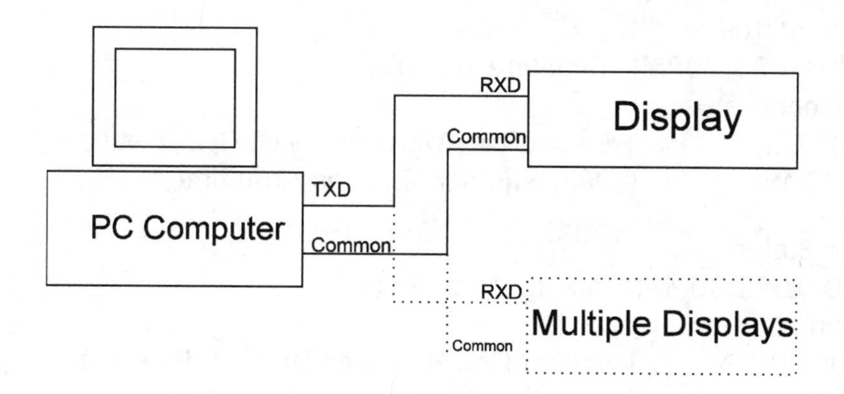

Figure 12.02

which can be placed at any desired cursor position in the display. A cursor-home command places the cursor at the upper left-hand corner of the display or at the very first character. With a special control character, the cursor can be configured to be visible or invisible.

In order to minimize the software requirements of this task, all 80 characters will be transmitted with each transaction, along with any required control characters. This display unit is configured for a data-byte frame of a Start bit, seven Data bits, Odd Parity, and one Stop bit for a total of 10 bits. At 9600 Baud, the transmission time for 80 display bytes and five control characters is approximately 88.5 milliseconds.

The home-monitor data will be displayed in page fashion. The pages will change every 10 seconds. Seven pages will output the data in the following manner.

Page 1: Outside temperature
  Line 1 [28Apr2000 01:23:00PM]   Real-time readout
  Line 2 [Outside Temperature   ]
  Line 3 [66.234   Deg. F   ]
  Line 4 [Leonik Engineering   ]

Page 2: First Floor Status
  Line 1 [28Apr2000 01:23:10PM] Real-time readout
  Line 2 [First Floor Status   ]
  Line 3 [Off   70.001 Deg. F   ] Status of the HVAC unit on or off /Temperature
  Line 4 [2 Cycles 60 Min   ] Daily Number of cycles / run time

```
Page 3:  Second Floor Status
  Line 1 [28Apr2000 01:23:20PM]  Real-time readout
  Line 2 [  Second Floor Status    ]
  Line 3 [On    70.001 Deg. F         ] Status of the HVAC unit on or off /Temperature
  Line 4 [ 2 Cycles  40 Min           ] Daily Number of cycles / run time

Page 4:  Water Pump Status
  Line 1 [28Apr2000 01:23:30PM]  Real-time readout
  Line 2 [  Water Pump Status     ]
  Line 3 [On     53.051 Psi           ] Status of the Pump unit on or off /Pressure
  Line 4 [8 Cycles   20 Min           ] Daily Number of cycles / run time

Page 5:  CrawlSpace temperature
  Line 1 [28Apr2000 01:23:40PM]  Real-time readout
  Line 2 [CrawlSpace Temperature]
  Line 3 [  55.123       Deg. F     ]
  Line 4 [  Leonik Engineering      ]  advertising

Page 6:  GreatRoom  Temperature
  Line 1 [28Apr2000  01:23:50PM]  Real-time readout
  Line 2 [GreatRoom Temperature ]
  Line 3 [  75.123       Deg. F     ]
  Line 4 [  Leonik Engineering      ]  advertising

Page 7:  Advertising
  Line 1 [28Apr2000 01:24:00PM ]  Real-time readout
  Line 2 [   VB 6  Real  World ....   ]
  Line 3 [          SAMS              ]
  Line 4 [Technical Publishing        ]
```

Now, if a door event should happen to occur, the door-event text will be immediately displayed for 10 seconds. The message will indicate the nature of the door event, such as "Front Door Open." Real time will not be displayed on Line 1. This task will be implemented with a first in/first out queue function identical to the wave-file queue used in Chapter 11. Once the door-event queue is depleted, the normal data pages will be displayed starting with Page 1.

The Code

The following code implements the desired display function. The first step is to add the required public variables and display constants to the end of the General Declaration section of Module1.bas. Fixed string variables of 20 bytes are specified for storage of the required display text. The use of these fixed variables eliminates problems of the lines being too short or too long and affecting the registration of the character placement in the display.

Add a form to the project and keep the default name of Form3. The new display-driver code will be placed in this form's code section.

Module1.bas - General Declaration Section – Add the following

```
' DISPLAY CONSTANTS

Public Const DisplayNO_CURSOR As Byte = 14
Public Const DisplayCURSOR As Byte = 15
Public Const DisplayCURSOR_HOME As Byte = 22
Public Const DisplayCLEAR As Byte = 21
Public Const DisplayHORIZ_SCROLL As Byte = 19
Public Const DisplayRESET As Byte = 20
Public Const DisplayESC As Byte = 25
Public Const DisplayBRIGHT As Byte = 79
Public Const DisplayDIM As Byte = 76

' DISPLAY Variables

Public vDisplayDoorQueue
Public sDisplayHeader As String
'fixed string variables specified to facilitate troublefree interfacing to the Display
Public sLine1 As String * 20
Public sLine2 As String * 20
Public sLine3 As String * 20
Public sLine4 As String * 20
```

Add the following statement to the Form1_Load event in order to activate the Form3 code. Form3.Visible = False.

The display-door events will use the same number structure that was used to launch the appropriate wave files.

1) Mailbox open event
2) Front door opened event
3) Front door closed event
4) Front door bell pushed event
5) Rear door opened event
6) Rear door closed event
7) Rear door bell pushed event

Make the following changes illustrated in **bold** to the following Sub procedures of Form2.

Form2 Sub Procedure – ShowFlags – Add the statements shown in **Bold**

```
If FrontDoorBellflag = True Then
lblFDB.BackColor = QBColor(10)
lblFDB.Caption = " Front Door Bell Switch On  "
lblFDbell.Caption = "Bell Rung: " & Now
 If tmrFDB.Enabled = False Then
 vWaveFileQue = vWaveFileQue  & "4"
 vDisplayDoorQueue = vDisplayDoorQueue & "4"
 tmrFDB.Enabled = True
 LogFrontDoorBell
 End If
Else: lblFDB.BackColor = QBColor(7)
lblFDB.Caption = " Front Door Bell Switch Off "
End If

If RearDoorBellflag = True Then
lblRDB.BackColor = QBColor(10)
lblRDB.Caption = " Rear Door Bell Switch On  "
lblRDbell.Caption = "Bell Rung: " & Now
```

If tmrRDB.Enabled = False Then
vWaveFileQue = vWaveFileQue & "7"
vDisplayDoorQueue = vDisplayDoorQueue & "7"
tmrRDB.Enabled = True
LogRearDoorBell
End If
Else: lblRDB.BackColor = QBColor(7)
lblRDB.Caption = " Rear Door Bell Switch Off "
End If

Form2 Sub Procedure CheckDoorChange – Add the statements shown in **Bold**

If FrontDoorOpenflag <> LastFrontDoorOpenflag Then
LastFrontDoorOpenflag = FrontDoorOpenflag
If FrontDoorOpenflag = True Then
vWaveFileQue = vWaveFileQue & "2"
vDisplayDoorQueue = vDisplayDoorQueue & "2"
lblFDopened.Caption = " Opened: " & Now
lblFDclosed.Caption = " Closed: ?"
Else:
vWaveFileQue = vWaveFileQue & "3"
vDisplayDoorQueue = vDisplayDoorQueue & "3"
lblFDclosed.Caption = " Closed: " & Now
End If
LogFrontDoor *' Time/Date Stamp the event*
End If

If RearDoorOpenflag <> LastRearDoorOpenflag Then
LastRearDoorOpenflag = RearDoorOpenflag
If RearDoorOpenflag = True Then
vWaveFileQue = vWaveFileQue & "5"
vDisplayDoorQueue = vDisplayDoorQueue & "5"
lblRDopened.Caption = " Opened: " & Now
lblRDclosed.Caption = " Closed: ?"
Else:

```
    vWaveFileQue = vWaveFileQue & "6"
    vDisplayDoorQueue = vDisplayDoorQueue & "6"
      lblRDclosed.Caption = " Closed: " & Now
    End If
    LogRearDoor ' Time/Date Stamp the event
  End If

  If MailBoxflag <> LastMailBoxflag Then
     LastMailBoxflag = MailBoxflag
    If MailBoxflag = True Then
    vWaveFileQue = vWaveFileQue & "1"
    vDisplayDoorQueue = vDisplayDoorQueue & "1"
      lblMBopened.Caption = " Opened: " & Now
      lblMBclosed.Caption = " Closed: ?"
    Else:
       lblMBclosed.Caption = " Closed: " & Now
    End If
    LogMailBoxDoor ' Time/Date Stamp the event
  End If
```

This next section of code is all new and is placed in the code section of Form3. In addition, add an MSComm object to the form called "MsComm1" and a timer to the form called "tmrDisplay".

Form3 Code Section – New Code

```
Option Explicit
```

```
Private Sub Form_Load()

 MSComm1.CommPort = 1 'choose an available Port
```

```
    ' 9600 baud, Odd parity, 7 data, and 1 stop bit.
    MSComm1.Settings = "9600,0,7,1"
    ' Open the port.
        MSComm1.PortOpen = True

   'Initialize the display
    sDisplayHeader = Chr(DisplayRESET)

   MSComm1.Output = sDisplayHeader 'Reset the Display
     MSComm1.Output = Time$ + "  " + Date$
  tmrDisplay.Interval = 300 'milliseconds
   sDisplayHeader = Chr(DisplayCURSOR_HOME) + Chr(DisplayNO_CURSOR) + _
       Chr(DisplayHORIZ_SCROLL) + Chr(DisplayESC) + Chr(DisplayBRIGHT)
  End Sub
```

The DisplayHORIZ_SCROLL constant instructs the display to allow the last line to scroll to the left after it is filled. This constant also prevents the display from scrolling vertically when the last line is filled. The display-brilliance control requires that you precede the desired brilliance with an escape character (ESC).

The Timer "tmrDisplay" is the driving force behind the alphanumeric display. The timer's interval is set to 300 milliseconds by the Form_Load event. Therefore, the timer's event Sub procedure is executed roughly three times per second. Essentially, a new second is determined by reading the system's real-time clock with a Now keyword statement. If the current Now result differs from the last recorded Now, a new second of time has elapsed.

A static variable called "iSecondsCounter" is incremented by one. If the value of this variable is greater than or equal to 10, a variable that points to the active displayed-data page is changed to the next page and the "iSecondsCounter" variable is set to zero. If the variable is less than 10, the time and date for the current display page is refreshed.

When a door event occurs, as determined by the state of the variable "vDisplayDoorQueue" not being a null, the variable " iSecondsCounter" is set to zero and the appropriate door-event page is displayed for a period of 10 seconds.

At the end of the 10-second period, if “vDisplayDoorQueue” is not empty, the next door-event page is transmitted to the display. This cycle continues until “vDisplayDoorQueue” is a zero-length string. When “vDisplayDoorQueue” is a zero-length string, the normal data pages will be displayed.

When writing the text to be displayed, it's not necessary to check the line lengths, because the fixed-string setting of the line string is set to 20 bytes. Any bytes over 20 are ignored, and any subsequent lines aren't affected.

```
Private Sub tmrDisplay_Timer()
'Define some local variables
Static vLastTime
Static iSecondsCounter As Integer
Static iDisplayScreenPointer As Integer

If Now <> vLastTime Then
  vLastTime = Now

  iSecondsCounter = iSecondsCounter + 1 'increment by one second
  If iSecondsCounter >= 10 Then
    iSecondsCounter = 0
    iDisplayScreenPointer = iDisplayScreenPointer + 1
    If iDisplayScreenPointer >= 8 Then iDisplayScreenPointer = 1
  End If

  If vDisplayDoorQueue <> “” And iDisplayScreenPointer <> 0 Then
  iSecondsCounter = 0
  iDisplayScreenPointer = 0
  End If

If iDisplayScreenPointer <> 0 Then
'Normal non door event path
sLine1 = Format(Date$, “ddmmmyyyy”) & “ “ & Format(Time$, “hh:mm:ssAM/PM”)

    Select Case iDisplayScreenPointer
    Case 1
```

```
    ShowOutsideTemp
   Case 2
    ShowFirstFloor
   Case 3
    ShowSecondFloor
   Case 4
    ShowWaterPump
   Case 5
    ShowCrawlSpaceTemp
   Case 6
    ShowGreatRoomTemp
   Case Else
     sLine2 = "   VB6 Real World....."
     sLine3 = "         SAMS        "
     sLine4 = "Technical Publishing"
   End Select
   Else:
     'Reset display seconds to provide equal time and
     'execute display door status
     If iSecondsCounter = 0 Then DisplayDoorStatus
 End If

     'Send to Display
     MSComm1.Output = sDisplayHeader
     MSComm1.Output = sLine1 & sLine2 & sLine3 & sLine4
End If
End Sub
```

```
Public Sub ShowOutsideTemp()

sLine2 = " Outside Temperature       "
sLine3 = "  " & vOutsideTemp & "   Deg. F  "
sLine4 = " LEONIK ENGINEERING "

End Sub
```

```
Public Sub ShowFirstFloor()
Dim s As String
If FirstFloorHvacOnflag = True Then s = " On" Else s = "Off"
sLine2 = " First Floor Status "
sLine3 = s & "   " & vFirstFloorTemp & " Deg. F"
sLine4 = lHvac1DailyCycle & " Cycles  " & lHvac1DailyOnTime & " Min.  "

End Sub
```

```
Public Sub ShowSecondFloor()
Dim s As String
If SecondFloorHvacOnflag = True Then s = " On" Else s = "Off"
sLine2 = " Second Floor Status      "
sLine3 = s & "   " & vSecondFloorTemp & " Deg. F"
sLine4 = lHvac2DailyCycle & " Cycles  " & lHvac2DailyOnTime & " Min.  "

End Sub
```

```
Public Sub ShowWaterPump()
Dim s As String
If WaterPumpOnflag = True Then s = " On" Else s = "Off"
sLine2 = " Water Pump Status      "
sLine3 = s & "    " & vWaterPressure & " Psi        "
sLine4 = lWaterPumpDailyCycle & " Cycles " & (lWaterPumpDailyOnTime /
    10)&_ " Min.  "

End Sub
```

```
Public Sub ShowCrawlSpaceTemp()

sLine2 = " CrawlSpace Temp.      "
sLine3 = "  " & vCrawlSpaceTemp & "   Deg. F  "
sLine4 = " LEONIK ENGINEERING "

End Sub
```

```
Public Sub ShowGreatRoomTemp()
```

```
sLine2 = " GreatRoom Temp.      "
sLine3 = "  " & vGreatRoomTemp & "   Deg. F  "
sLine4 = " LEONIK ENGINEERING "

End Sub
```

```
Public Sub DisplayDoorStatus()
   Dim DoorDisplay
   'Remove left most character and place the remainder back into
   'the variable.
   DoorDisplay = Mid$(vDisplayDoorQueue, 1, 1)
vDisplayDoorQueue=Mid$(vDisplayDoorQueue,2, Len(vDisplayDoorQueue))
Select Case DoorDisplay

Case 1
   ShowMailBox
Case 2
   ShowFrontDoorOpen
Case 3
   ShowFrontDoorClosed
Case 4
   ShowFrontDoorBell
Case 5
   ShowRearDoorOpen
Case 6
   ShowRearDoorClosed
Case 7
   ShowRearDoorBell
Case Else
End Select

End Sub
```

```
Public Sub ShowMailBox()
sLine1 = "      The Mail      "
```

```
sLine2 = "                "
sLine3 = "      is        "
sLine4 = "     Here !!!!    "

End Sub
```

```
Public Sub ShowFrontDoorOpen()
sLine1 = "       The      "
sLine2 = "    Front Door       "
sLine3 = "       Just     "
sLine4 = "     Opened      "

End Sub
```

```
Public Sub ShowFrontDoorClosed()
sLine1 = "       The      "
sLine2 = "    Front Door    "
sLine3 = "       Just      "
sLine4 = "     Closed      "

End Sub
```

```
Public Sub ShowFrontDoorBell()
sLine1 = "       The       "
sLine2 = "  Front Door Bell   "
sLine3 = "                "
sLine4 = "                "

End Sub
```

```
Public Sub ShowRearDoorOpen()
sLine1 = "       The       "
sLine2 = "    Rear Door     "
sLine3 = "       Just      "
sLine4 = "     Opened      "
```

```
End Sub
```

```
Public Sub ShowRearDoorClosed()
sLine1 = "     The      "
sLine2 = "  Rear Door    "
sLine3 = "     Just      "
sLine4 = "    Closed     "

End Sub
```

```
Public Sub ShowRearDoorBell()
sLine1 = "     The      "
sLine2 = " Rear Door Bell  "
sLine3 = "              "
sLine4 = " Rear Door Bell  "

End Sub
```

Conclusion

The remote display serial driver function was added to the home-monitor project with ease and a small amount of coding. This feature allows you to place any number of display modules in remote locations to provide a readout of all the current data collected by the home-monitor software, including door events. This chapter ends with a sampling of actual alphanumeric display pages created by the software in this chapter. Figure 12.03 shows an unpackaged display.

Sample Display Pages

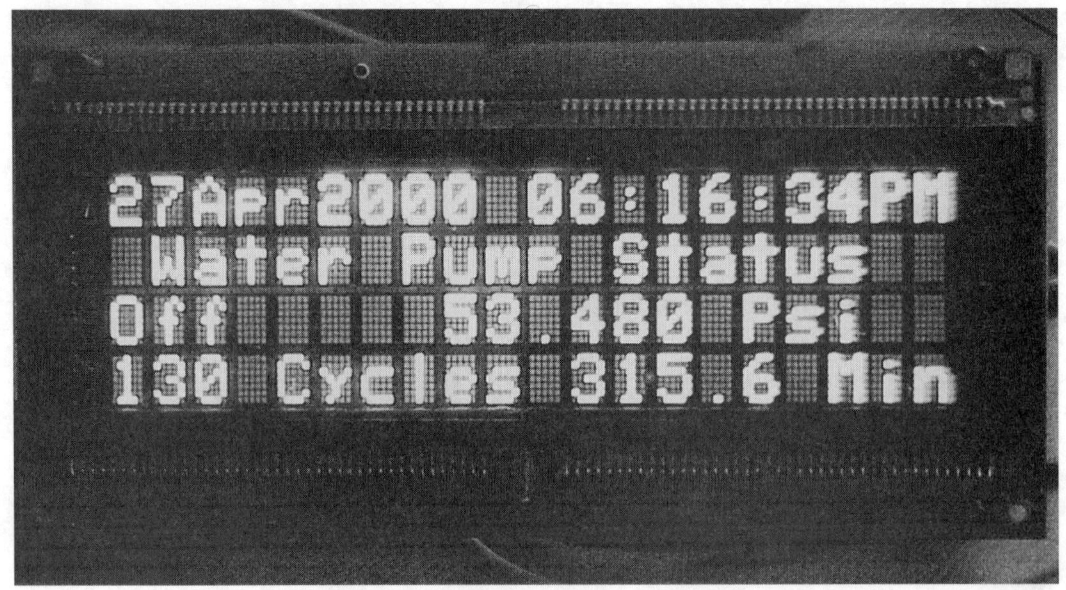

Figure 12.03

Figure 12.04

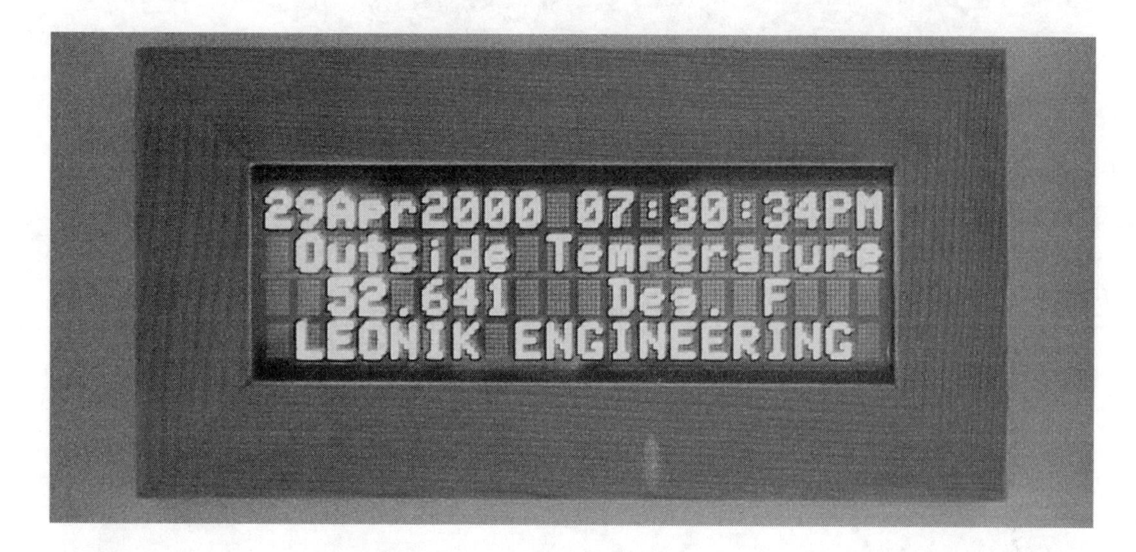

Figure 12.05

Figure 12.06

29Apr2000 07:43:47PM
First Floor Status
Off 69.161 Deg. F
27 Cycles 186 Min.

Figure 12.07

Chapter 13
CONTROL

INTRODUCTION

This chapter explores the implementation of a very simple control scheme. The home-monitor house has air-supply ductwork located near the top of the cathedral ceiling in the great room. This ductwork feeds a blower that moves the hot air trapped by the cathedral ceiling to a cooler section of the house. If the fireplace is on, or if there's a significant heat gain from solar radiation through the windows, the hot air will rise into the cathedral ceiling. The great room temperature sensor is located at the ductwork level and is already being monitored by the home-monitor program. The home-monitor program will control PLC output "O/1" based on the measured temperature. The PLC output drives a solid-state relay, which controls whether the blower is on or off.

Task Description

A simple control scheme will be used to implement the blower control. The blower can be manually turned on or off by double-clicking on the Form2 label "lblO1Cmd". An automatic control scheme will be installed that is enabled or disabled by a checkbox. If the automatic control checkbox is enabled, the blower will automatically turn on when the great room temperature is greater than 72.5 degrees Fahrenheit. Once the blower is turned on, it will remain on until the great room temperature drops below the dead-band specification. The dead-band specification in this example is fixed at a value of three degrees Fahrenheit.

Hysteresis, sometimes called dead band, is control terminology. This example is a simple two-state control system, or On/Off control. In two-state systems, hysteresis is required to prevent the output from rapidly cycling on and off around the starting point. This rapid cycling on and off can easily burn out a motor. A typical water pump that supplies water to a house will have a hysteresis value of 20 psi. The well-water pump usually turns on at a value of 35 psi and turns off at a pressure value of 55 psi, without an acceptable level of hysteresis.

Therefore, this control system with a hysteresis value of three degrees Fahrenheit only needs to know the starting temperature point for the blower. The blower-stop temperature point in this application is determined by subtracting the hysteresis value from the blower-start temperature point. You want the blower to turn on and convey the hot air to a cooler section of the house. Once the blower is on, you want it to turn off when the temperature drops below a certain value. This is the opposite control action of the well-water pump.

Form3 will be used as the control page. Change the caption of Form3 to read "GreatRoom Cathedral Ceiling Blower Control". The proposed look for Form3 is depicted in Figure 13.01. Add the necessary components to your Form. The runtime-controlled labels all have an "lbl" prefix. The rectangular boxes on the form are not controlled at run time and are simply rectangular shapes. The lines feeding the rectangular shapes and the blower picture box will be animated with color at run time. In addition, the fan image in the blower picture box will appear to be rotating when the blower is running.

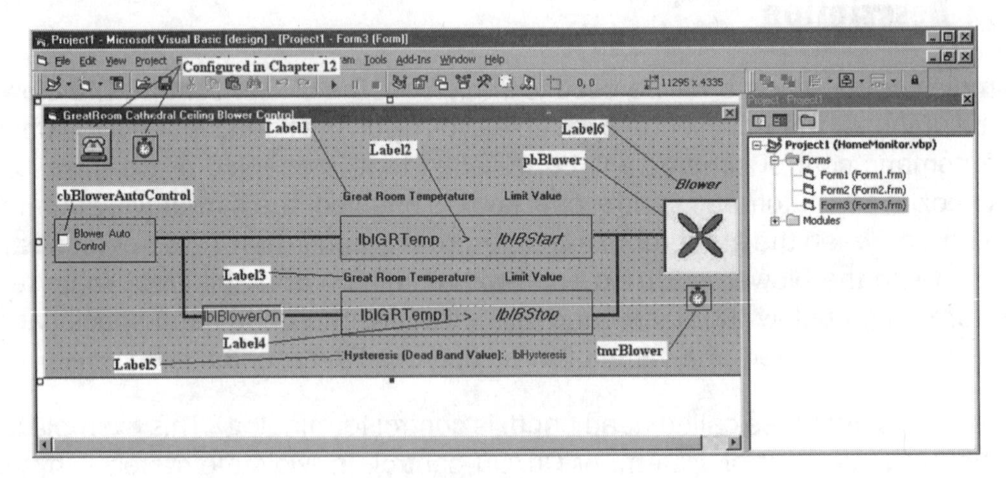

Figure 13.01

The timer "tmrBlower" controls which image is placed in the picture box "pbBlower". A 100-millisecond time interval is specified for this timer. The fan images were created with a CAD program and saved as a JPEG image file. The file names are fan11.jpg and fan21.jpg. These file are provided in "Chapter 13" on the companion CD-ROM. Figure 13.02 shows the graphic of fan11.jpg. The graphic for fan21.jpg is shown in Figure 13.03. Whenever the fan is running—whether it's in manual mode or automatic mode—the blower timer will be enabled. The image in the blower picture box will change with each blower-timer event, thereby creating the illusion of rotation.

Figure 13.02 **Figure 13.03**

The illustration in Figure 13.04 provides identification to the various runtime-controlled line components. All the lines are configured as an array of Line1. The numbers listed in Figure 13.04 represent the index number of Line1. These index numbers are provided because specific lines will change colors based upon the active control path.

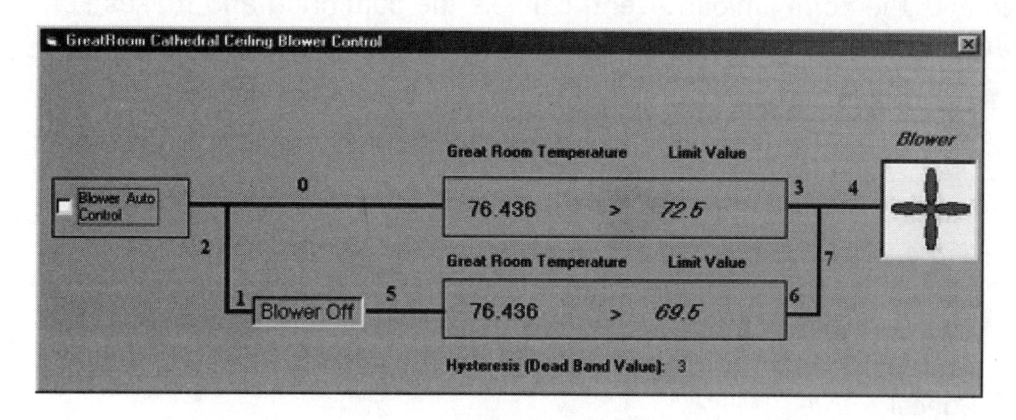

Figure 13.04

The Code

Form2 will require some minor changes to implement this control function. You now need the ability to view Form3. Add a checkbox to Form2 just a little to the right of the "View Communication Window" checkbox. The name of the new checkbox is "cbViewControlScreen". Change the new checkbox's caption to read "View Control Screen". Add a label directly over the label "lblO1Cmd". The name of the new label is "lblBlowerAuto". Change the new label's caption to read "Auto Control". This label will indicate whether the great room blower is on automatic control.

Form2 Add cbViewControlScreen Sub Procedure

```
Private Sub cbViewControlScreen_Click()
If cbViewControlScreen.Value = 1 Then
 Form3.Visible = True
Else: Form3.Visible = False
End If
End Sub
```

The Form_QueryUnload Sub procedure ensures that Form3 is always left running the control task. If you click on the Form3 exit box, Form_QueryUnload intercepts the Form_unload event, cancels the command and makes Form3 no longer visible.

Form3 Add the following Sub Procedures

```
Private Sub Form_QueryUnload(Cancel As Integer, UnloadMode As Integer)
'if the visible form is told to close then make the form not visible
'Cancel the close command and change the state of the checkbox on form2
Form3.Visible = False
Form2.cbViewControlScreen.Value = 0
Cancel = True
End Sub
```

```
Public Sub BlowerControl()

'show GreatRoom Temperature
lblGRTemp.Caption = vGreatRoomTemp
lblGRTemp1.Caption = vGreatRoomTemp

If cbBlowerAutoControl.Value = 1 Then
  Line1(0).BorderColor = QBColor(10) 'Fill with Green
  Line1(1).BorderColor = QBColor(10)  'Fill with Green
  Line1(2).BorderColor = QBColor(10)  'Fill with Green

  If Status01flag = True Then BlowerOnFlag = True
     ' Process Control
   If BlowerOnFlag = False Then
   If Val(vGreatRoomTemp) > vBlowerStart Then BlowerOnFlag = True
   Else
   If Val(vGreatRoomTemp) < vBlowerStop Then BlowerOnFlag = False
   Out1CmdFlag = False 'allow control only when cbBlowerAutoControl is true
   ' this allows Manual control on Form2 0/1
   End If
  Form2.lblBlowerAuto.Visible = True
Else:
  Line1(0).BorderColor = QBColor(0) 'Fill with Black
  Line1(1).BorderColor = QBColor(0) 'Fill with Black
  Line1(2).BorderColor = QBColor(0) 'Fill with Black
  BlowerOnFlag = False
  Form2.lblBlowerAuto.Visible = False
  End If

'This statement paints line1(3) black when the temp drops below the start point
If Val(vGreatRoomTemp) > vBlowerStart And cbBlowerAutoControl.Value = 1 Then
Line1(3).BorderColor = QBColor(10) 'Fill with Green
Else: Line1(3).BorderColor = QBColor(0) 'Fill with Black
End If
```

```
If BlowerOnFlag = True Then
 lblBlowerOn.Caption = " Blower ON "
 lblBlowerOn.BackColor = QBColor(10) 'Fill with Green
 Out1CmdFlag = True
 Line1(4).BorderColor = QBColor(10) 'Fill with Green
 Line1(5).BorderColor = QBColor(10) 'Fill with Green
 Line1(6).BorderColor = QBColor(10) 'Fill with Green
 Line1(7).BorderColor = QBColor(10) 'Fill with Green
Else:
 lblBlowerOn.Caption = " Blower Off"
 lblBlowerOn.BackColor = QBColor(7) 'Fill with Gray
 Line1(4).BorderColor = QBColor(0) 'Fill with Black
 Line1(5).BorderColor = QBColor(0) 'Fill with Black
 Line1(6).BorderColor = QBColor(0) 'Fill with Black
 Line1(7).BorderColor = QBColor(0) 'Fill with Black
End If
'show true fan status in auto control or manual control
If Status01flag = True Then
tmrBlower.Enabled = True
Else: tmrBlower.Enabled = False
End If
End Sub
```

```
Private Sub tmrBlower_Timer()
' A alternation scheme utilizing a Tag property
If pbBlower.Tag = "" Then
pbBlower.Picture = LoadPicture("fan21.jpg")
pbBlower.Tag = "Next Frame"
Else: pbBlower.Picture = LoadPicture("fan11.jpg")
  pbBlower.Tag = ""
End If
End Sub
```

The following three illustrations show the line-coloring based upon the active control paths. In Figure 13.05, the great room temperature is greater than all

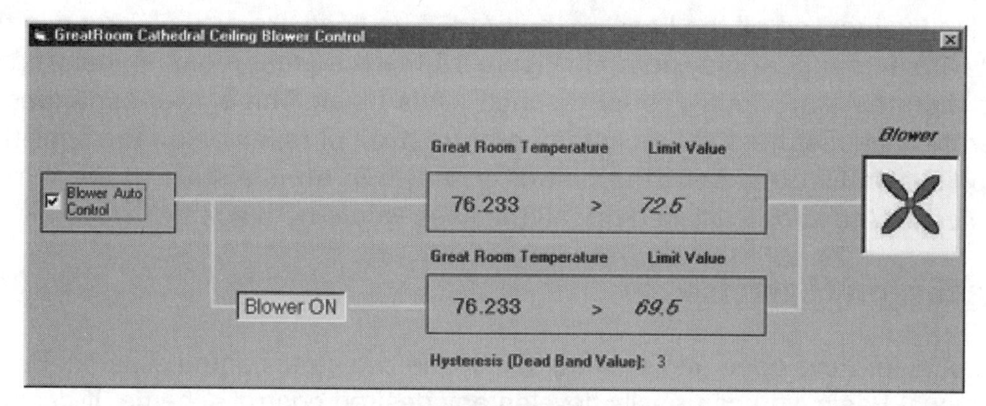

Figure 13.05

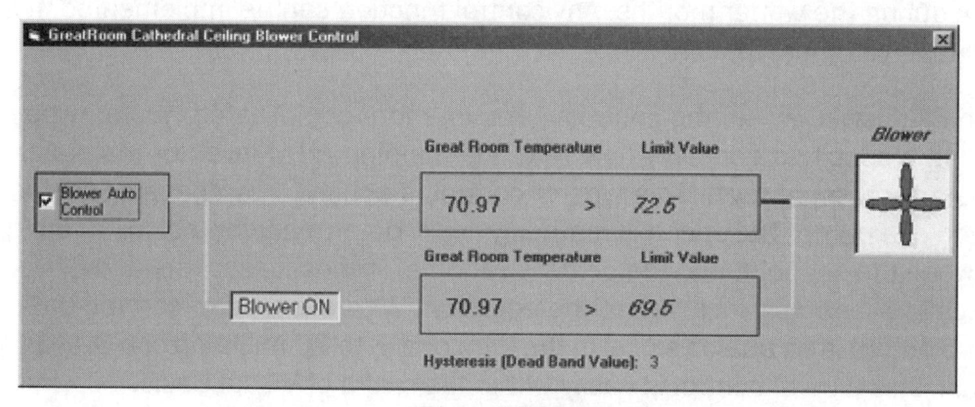

Figure 13.06

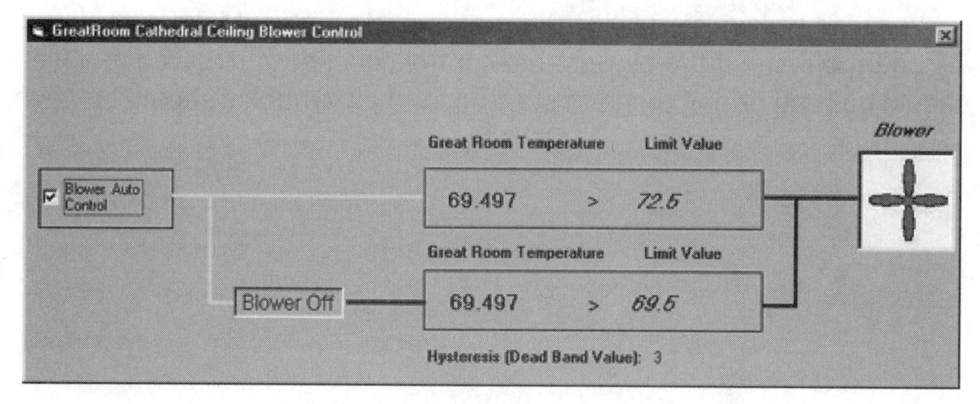

Figure 13.07

the limit values, so all the lines are colored. (The lines are actually green in the software, but appear gray here.) In Figure 13.06, the great room temperature is less than the start limit. Therefore, line1 (3) is black. The blower continues to operate because the great room temperature has not fallen below the specified stop limit. In Figure 13.07, the blower is shown in an idle state. If the blower auto-control textbox is clicked off, all the lines would be black.

Conclusion/Exercise

This chapter developed and animated a simple two-state control scheme. By using Visual Basic, you can easily develop any desired control scheme. If desired, you can include time in the logic equation—or perhaps the control should only work during the winter months. Any control function can be implemented. It's really up to your imagination.

In the industrial world, the controls get a lot more complicated. You may be controlling a specified constant flow rate, maintaining a PH level, or maintaining a constant pressure level. This type of control is achieved by changing the actual speed of a pump. The speed is mathematically determined by looking at the error between the set point and the process variable. In most cases, the derivative, the integral, or both are included in the equation. With this type of control, the command output is an analog signal in the form of a 4- to 20-milliamp or a 0- to 10-volt signal. This type of control is easily within the realm of Visual Basic.

As an exercise, you should change the program so that the start limit and hysteresis values can be adjusted while the program is running. Make sure that these values and the state of the blower auto-control checkbox are saved to a file. This file should be read during program start-up, and all variables should be restored.

Chapter 14
CONCLUSION

Initially, when I thought about an application example for this book, a dozen industrial-type applications rushed through my head. I finally settled on a home-monitor application, since I felt it had a broader appeal and would be easier to comprehend. I'm glad I did, because I had a lot of fun writing the software and writing this book. However, the material covered in this book can be applied easily to any industrial-based application.

Actually, the book just scratches the surface of the types of animation and control that can be implemented using Visual Basic 6. Many commercially available controls can be added to enhance your Visual Basic projects. In addition, Visual Basic allows you to create your own specialized controls.

One area that this book doesn't address is the Web. Perhaps this should be the topic of my next book, "VB6 Real-World Interfacing, Animation and Control: The Web." Once you have the information in the computer, virtually anything is possible. Website information can be updated. Email alerts can be issued. The alphanumeric display software module can be programmed to periodically provide real-time quotes of your favorite stocks. This information can be scrolled across the display—or displays, for that matter. It's really up to your imagination.

We live in truly exciting times. Today, software has evolved to an incredible level. Extremely powerful computers can be purchased for under $1000. The Web provides instant information on just about any subject. If you're knowledgeable in electronics, you can design a circuit with schematic-capture software, link the design's netlist to a printed circuit-board CAD software package, email the files to a printed circuit-board fabrication house, and then, two days later, the circuit boards arrive at your doorstep. Finally, software such as Visual Basic 6 provides the bridge between the circuit board and the PC world.

I hope this book has accomplished its task of introducing you to the power of Visual Basic through the application of interfacing to real-world devices. The software presented here was kept simple deliberately, in order to clearly convey a ground-level approach. There are many things that can be done to enhance the software and give the animation a more polished look. More colors could be used in the creation of the screens. Multiple screens could be used to present the information. Additional features and information can be provided. True to its title, this book has provided you with the basics; the enhancements, then, are up to you.

APPENDIX

A NOTE ON THE SOFTWARE

Most of the software was constructed with a display size of 1024 x 768. The software for the home-monitor project maintains the software progression of each chapter. For example, the software contained in Chapter 10 builds on the software contained in Chapter 9.

For best results, copy the folder from the CD-ROM into the root directory of your hard drive. Then use Explorer to run the program in its respective folder. If you want to modify the program but the system is locked, one solution is to use Explorer to change the property setting of all the Visual Basic files to not be "Read Only."

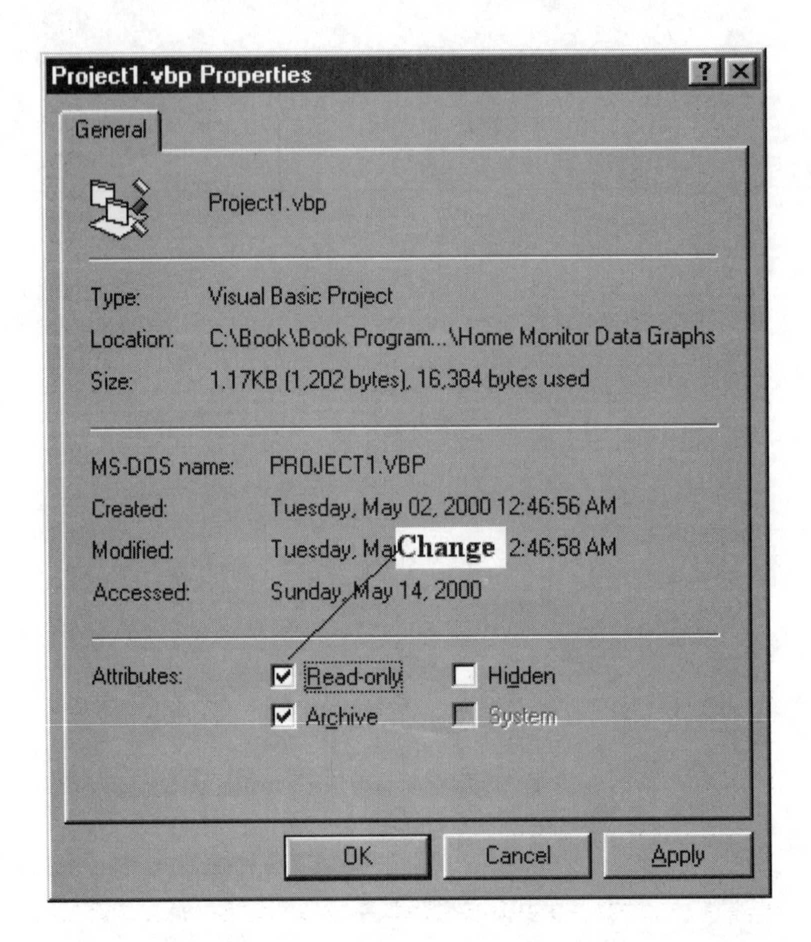

ASCII TABLE

HEX	DECIMAL	CHAR.	DESCRIPTION	HEX	DECIMAL	CHAR.	DESCRIPTION
00	00	NUL	Null	40	64	@	Printable Character
01	01	SOH	Start of Header	41	65	A	Printable Character
02	02	STX	Start of Text	42	66	B	Printable Character
03	03	ETX	End of Text	43	67	C	Printable Character
04	04	EOT	End of Transmission	44	68	D	Printable Character
05	05	ENQ	Enquire	45	69	E	Printable Character
06	06	ACK	Acknowledge	46	70	F	Printable Character
07	07	BEL	Bell	47	71	G	Printable Character
08	08	BS	BackSpace	48	72	H	Printable Character
09	09	HT	Horizontal Tab	49	73	I	Printable Character
OA	10	LF	Line Feed	4A	74	J	Printable Character
0B	11	VT	Vertical Tab	4B	75	K	Printable Character
0C	12	FF	Form Feed	4C	76	L	Printable Character
0D	13	CR	Carriage Return	4D	77	M	Printable Character
0E	14	SO	Shift Out	4E	78	N	Printable Character
0F	15	SI	Shift In	4F	79	Q	Printable Character
10	16	DLE	Data Link Escape	50	80	P	Printable Character
11	17	DC1	Device Control 1	51	81	Q	Printable Character
12	18	DC2	Device Control 2	52	82	R	Printable Character
13	19	DC3	Device Control 3	53	83	S	Printable Character
14	20	DC4	Device Control 4	54	84	T	Printable Character
15	21	NAK	Negative acknowledge	55	85	U	Printable Character
16	22	SYN	Synchronous Idle	56	86	V	Printable Character
17	23	ETB	End of Transmission Block	57	87	W	Printable Character
18	24	CAN	Cancel	58	88	X	Printable Character
19	25	EM	End of Medium	59	89	Y	Printable Character
1A	26	SUB	Substitute	5A	90	Z	Printable Character
1B	27	ESC	Escape	5B	91	[	Printable Character
1C	28	FS	Field Separator	5C	92	\	Printable Character
1D	29	GS	Group Separator	5D	93	]	Printable Character
1E	30	RS	Record Separator	5E	94	^	Printable Character
1F	31	US	Unit Separator	5F	95	_	Printable Character
20	32	SP	Printable Character	60	96	(	Printable Character
21	33	!	Printable Character	61	97	a	Printable Character
22	34	“	Printable Character	62	98	b	Printable Character
23	35	#	Printable Character	63	99	c	Printable Character
24	36	$	Printable Character	64	100	d	Printable Character
25	37	%	Printable Character	65	101	e	Printable Character
26	38	&	Printable Character	66	102	f	Printable Character
27	39	(	Printable Character	67	103	g	Printable Character
28	40	(	Printable Character	68	104	h	Printable Character
29	41	)	Printable Character	69	105	i	Printable Character
2A	42	*	Printable Character	6A	106	j	Printable Character

Continued from previous page

HEX	DECIMAL	CHAR.	DESCRIPTION	HEX	DECIMAL	CHAR.	DESCRIPTION
2B	43	+	Printable Character	6B	107	k	Printable Character
2C	44	,	Printable Character	6C	108	l	Printable Character
2D	45	-	Printable Character	6D	109	m	Printable Character
2E	46	.	Printable Character	6E	110	n	Printable Character
2F	47	/	Printable Character	6F	111	o	Printable Character
30	48	0	Printable Character	70	112	p	Printable Character
31	49	1	Printable Character	71	113	q	Printable Character
32	50	2	Printable Character	72	114	r	Printable Character
33	51	3	Printable Character	73	115	s	Printable Character
34	52	4	Printable Character	74	116	t	Printable Character
35	53	5	Printable Character	75	117	u	Printable Character
36	54	6	Printable Character	76	118	v	Printable Character
37	55	7	Printable Character	77	119	w	Printable Character
38	56	8	Printable Character	78	120	x	Printable Character
39	57	9	Printable Character	79	121	y	Printable Character
3A	58	:	Printable Character	7A	122	z	Printable Character
3B	59	;	Printable Character	7B	123	{	Printable Character
3C	60	<	Printable Character	7C	124	«	Printable Character
3D	61	=	Printable Character	7D	125	}	Printable Character
3E	62	>	Printable Character	7E	126	~	Printable Character
3F	63	?	Printable Character	7F	127	DEL	Delete

INDEX

A

B

C

D

E

F

G

H

I

J

K

L

M

N

O

P

Q

R

S

W

X